Introduction to Unified
Strength Theory

Introduction to Unified Strength Theory

Mao-Hong Yu

Xi'an Jiaotong University, China

Shu-Qi Yu

The University of British Columbia, Canada

CRC Press
Taylor & Francis Group
Boca Raton   London   New York   Leiden

CRC Press is an imprint of the
Taylor & Francis Group, an **informa** business

A BALKEMA BOOK

CRC Press/Balkema is an imprint of the Taylor & Francis Group, an informa business

© 2019 Taylor & Francis Group, London, UK

Typeset by Apex CoVantage, LLC

Library of Congress Cataloging-in-Publication Data
Applied for

Published by: CRC Press/Balkema
 Schipholweg 107c, 2316 XC Leiden,
 The Netherlands e-mail: Pub.NL@taylorandfrancis.com
 www.crcpress.com – www.taylorandfrancis.com

ISBN: 978-0-367-24682-2 (Hbk)
ISBN: 978-0-429-28394-9 (eBook)

Contents

Preface   viii

1   Introduction   1

   1.1   Strength of materials in tension and
       compression 1
   1.2   Strength of materials under complex stress 1
   1.3   Single strength theory 4
   1.4   Voigt-Timoshenko conundrum 7
   1.5   The development of unified strength theory 9
   1.6   Applications of the Yu-UST 10
   1.7   Summary 10
       References 12
       Readings 15

2   Stress state and elements   17

   2.1   Introduction 17
   2.2   Stresses on the oblique plane 19
   2.3   Cubic hexahedron and octahedron 20
   2.4   Mohr stress circle 23
   2.5   Stress space 24
   2.6   Summary 28
       References 28
       Readings 28

3   Yu-unified strength theory (Yu-UST)   31

   3.1   Introduction 31
   3.2   The Yu-unified strength theory 31

3.3 Special cases of the Yu-UST for different parameter b 34

3.4 Limit loci of the Yu-UST by varying parameter b in the π-plane 35

3.5 Variation of limit loci of the Yu-UST when $\alpha = 1/2$ 38

3.6 Limit surfaces of the Yu-UST in principal stress space 40

3.7 Limit loci of the Yu-UST in plane stress state 42

3.8 Limit loci of the Yu-UST under the $\sigma - \tau$ combined stress state 43

3.9 Experiment verification of Yu-UST 46

3.10 Summary 49

References 49

Readings 50

4 Yu-unified yield criteria for metals   **51**

4.1 Introduction 51

4.2 The Yu-unified yield criterion (Yu-UYC) 51

4.3 Special cases of the Yu-UYC 52

4.4 Experimental determination of parameter b 55

4.5 Yu-UYC in the plane stress state 56

4.6 Yu-UYC in the $\sigma - \tau$ stress state 58

4.7 Yield surfaces of the Yu-UYC 61

4.8 Summary 61

References 62

Readings 64

5 Applications of the Yu-UST and Yu-UYC   **65**

5.1 Introduction 65

5.2 Strength design of a thin-walled pressure cylinder 65

5.3 Thickness design of a thin-walled pressure vessel 68

5.4 Strength design of a gun tube of trench mortar 70

5.5 Strength design of an oil pipe 72

5.6 Strength design of an automobile transmission shaft: maximum allowable torque 73

5.7 Strength design of an automobile transmission shaft: outer diameter design of drive shaft 76

5.8 Strength design of a transmission axle 77

5.9 Unified solution for elastic limit pressure of a thick-walled cylinder subjected to internal pressure 79

5.10 Analysis of plastic ultimate pressure for a thick-walled cylinder with internal pressure 86

5.11 Unified solution of plastic limit internal pressure for a thick-walled cylinder with the same tensile strength and compressive strength materials 90

5.12 Unified solution of plastic limit internal pressure for a thick-walled cylinder with different tensile strength and compressive strength materials 91

5.13 Barrel strength design of artillery 94

5.14 The economic significance of the use of Yu-unified strength theory 96

5.15 Procedure of the applications of the Yu-UST and Yu-UYC 97

5.16 Summary 99

References 99

Readings 100

Preface

Strength theory is very important in solid mechanics and engineering. We need to study it in the course of mechanics of materials. It contains four or five classical strength theories. Sometimes strength theory is referred to as the strength hypothesis.

We also need to study strength theory in plasticity. It is known as yield criterion, yield condition or yield function.

Strength theory is one of the theoretical foundations in soil mechanics and rock mechanics. It is called the failure criterion.

We will use the strength theory in the course of machine parts design, strength of structure and others.

A lot of strength theories have been proposed. It is not easy to choose a reasonable strength theory in the strength design of structure. For example, the maximum principal strain theory is used for the strength design of barrel strength of cannons in Russia and China; however, maximum shear stress theory is adopted in Germany, and maximum distortion energy theory is used in the United States. Which one is the more reasonable choice?

The formula of strength theory is simple. The development of strength theory, however, is slow. The maximum shear stress theory used for non-SD (strength difference in tension and compression) material was proposed by Tresca in 1864. The expression of the Tresca yield criterion is

$$\sigma_1 - \sigma_3 = \sigma_y$$

The Mohr-Coulomb criterion used for SD material was proposed in 1900, 36 years after the Tresca yield criterion. The expression of the Mohr-Coulomb criterion is

$$\sigma_1 - \alpha\sigma_3 = \sigma_t, \alpha = \sigma_t / \sigma_c$$

It can be seen that the difference between them is only one symbol.

Early in 1901, professor Voigt conducted a lot of tests to check Mohr's strength theory at Göttingen University, Germany. These experimental results do not agree well with Mohr's theory. Therefore, Voigt concluded that strength theory is too complicated, and it is impossible to formulate a single strength criterion that can be applied to all kinds of structural materials. In 1953, Prof. Timoshenko at Stanford University repeated Voigt's conclusion that it is impossible to devise a single theory for successful application to all kinds of structural materials. In 1984, Bowes, Russell and Suter stated in *Mechanics of Engineering Materials* that "An understanding of failure theory is necessary in order to avoid making some serious errors. Unfortunately, no single theory will be found to apply in all cases." The same thought was expressed in the *Encyclopedia of China* that "it is impossible to establish a unified strength theory for various materials" (1985). This problem was referred to as the "Voigt-Timoshenko conundrum." This is a conundrum that related to strength theory in the 20th century, and it was unsolved until 1990.

The Drucker postulate (1951) provides a theoretical frame relating to the yield criterion. The convexity of yield surface for material under complex stress appeared as well. The study of the yield criterion may be developing on a more reliable theoretical basis. The lower bound and the upper bound of the yield surfaces of strength theory can be obtained by the Drucker postulate. The yield surfaces of an expected unified strength theory should cover the whole regions of convex strength theory from the lower bound to the upper bound.

Unified strength theory was proposed by Mao-Hong Yu in 1991. It was three decades from the twin-shear yield criterion (1961) to twin-shear strength theory (1985) and to the unified strength theory (1991). The mathematic expressions are shown as follows.

Twin-shear yield criterion for non-SD materials (Yu, 1961).

$$f = \sigma_1 - \frac{1}{2}(\sigma_2 + \sigma_3) = \sigma_t, \text{ when } \sigma_2 \leq \frac{1}{2}(\sigma_1 + \sigma_3)$$

$$f' = \frac{1}{2}(\sigma_1 + \sigma_2) - \sigma_3 = \sigma_t, \text{ when } \sigma_2 \geq \frac{1}{2}(\sigma_1 + \sigma_3)$$

Twin-shear strength theory for SD materials (Yu *et al.*, 1985)

$$F = \sigma_1 - \frac{\alpha}{2}(\sigma_2 + \sigma_3) = \sigma_t, \text{ when } \sigma_2 \leq \frac{\sigma_1 + \alpha\sigma_3}{1 + \alpha}$$

$$F' = \frac{1}{2}(\sigma_1 + \sigma_2) - \alpha\sigma_3 = \sigma_t, \text{ when } \sigma_2 \geq \frac{\sigma_1 + \alpha\sigma_3}{1 + \alpha}$$

Unified strength theory for various materials (Yu and He, 1991)

$$F = \sigma_1 - \frac{\alpha}{1+b}(b\sigma_2 + \sigma_3) = \sigma_t, \text{ when } \sigma_2 \leq \frac{\sigma_1 + \alpha\sigma_3}{1 + \alpha}$$

$$F' = \frac{1}{1+b}(\sigma_1 + b\sigma_2) - \alpha\sigma_3 = \sigma_t, \text{ when } \sigma_2 \geq \frac{\sigma_1 + \alpha\sigma_3}{1 + \alpha}$$

The limit surfaces of the unified strength theory cover all the convex regions from the inner bound to outer bound, as shown in Figure 0.1. No other strength theory can cover all the convex regions.

The mathematical expression of the unified strength theory is very simple. However, the unified strength theory is rich in content. It has been applied in many fields.

Systemic descriptions of the unified strength theory can be found in two books. The first was published in Chinese in 1992 titled *New System of Strength Theory*. The second was published in English titled *Unified Strength Theory and Its Applications* in 2004. These two books are suitable for researchers and engineers who are familiar with the plasticity, geo-mechanics and strength analyses of structure.

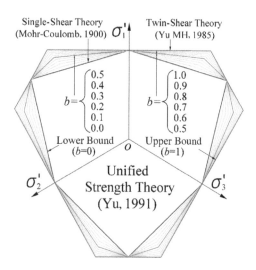

Figure 0.1 Limit loci of unified strength theory on the deviatoric plane

Now, we need a simple description about the unified strength theory for undergraduate students who are studying the mechanics of materials and engineering mechanics, as well as graduate students who are interested in this field. The unified strength theory has a simple mathematical expression and clear physical conception. It can be easily understood by undergraduate students and graduate students. Researchers and engineers can also benefit from this book.

Finally, the authors wish to acknowledge the support of Xi'an Jiaotong University Alumni Association of Hong Kong; Xi'an Jiaotong University Alumni Association of Civil Engineering Department; and State Key Laboratory for Strength and Vibration of Mechanical Structures, Xi'an Jiaotong University, Xi'an, China. We are also grateful to Jia-Yu Liang, the research assistant of Prof. Mao-Hong Yu, for his help in writing this book. I would like also to express my sincere thanks to the editors Alistair Bright and Umamaheswari Chelladurai for their excellent editorial work on this book.

Mao-Hong Yu and Shu-Qi Yu
Spring 2019

References

Yu, M.H. (1961) *General Behavior of Isotropic Yield Function.* Res. Report of Xi'an Jiaotong University, Xi'an, China (in Chinese).

Yu, M.H., He, L.N. & Song, L.Y. (1985) Twin shear stress theory and its generalization. *Science in China Series A, English Edition*, 28(11), 1174–1183.

Yu, M.H. & He, L.N. (1991) A new model and theory on yield and failure of materials under the complex stress state. In: Jono, M. and Inoue, T. (eds) *Mechanical Behavior of Materials-6 (ICM-6)*, Vol. 3. Pergamon Press, Oxford, pp. 841–846.

1 Introduction

1.1 Strength of materials in tension and compression

Strength theory deals with the yield or failure of materials under the complex stress state (combined stress state or multi-axial stress state). It is very important in solid mechanics and engineering. The uniaxial tensile and compressive strength of material can be obtained by simple test equipment. This kind of experimental equipment can be found at most engineering colleges and in many factories. The stress-strain curves of mild steel and iron in tension and compression are shown in Figures 1.1 and 1.2, respectively. The two figures show that there exist a tensile yield or failure point and a compressive yield or failure point at which the material will begin to deform plastically or failure. These two figures are well-known for students and engineers.

For most homogeneous, continuous and isotropic materials, the axial tensile strength (σ_t) is different from the axial compressive strength (σ_c). The yield stresses of some materials, such as mild steel, are identical both in tension and compression, as shown in Figure 1.1.

1.2 Strength of materials under complex stress

Most materials are acted upon under a complex stress state. If we take the tension as a positive and compression as a negative, the stress state may be presented as a stress point in a three-dimensional space. The stress point could be situated anywhere within the three-dimensional space of the principal stresses as shown in Figure 1.3. The stress point $P(\sigma_1, \sigma_2, \sigma_3)$ of different signs could combine up to eight quadrants of $(+++), (++-), (+-+), (+--), (-++), (-+-), (--+)$ and $(---)$.

For example, there is a material with axial tensile strength on point A $(\sigma_1 = \sigma_t, \sigma_2 = 0, \sigma_3 = 0)$ and axial compressive strength on point B $(\sigma_1 = -\sigma_c, \sigma_2 = 0, \sigma_3 = 0)$. It is seen that the axial tensile and compressive strengths are different. The strength of material under the general

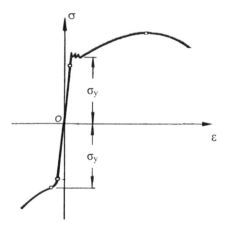

Figure 1.1 Stress-strain curve of steel

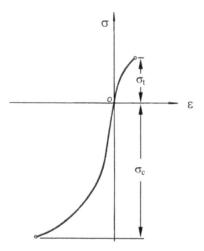

Figure 1.2 Stress-strain curve of iron

stress state, such as point P in Figure 1.3, is far more complicated than that in a simple stress state. The strength of materials under the complex stress state (bi-axial stress or tri-axial stress) is hard to determine because the experimental equipment is rare and very expensive.

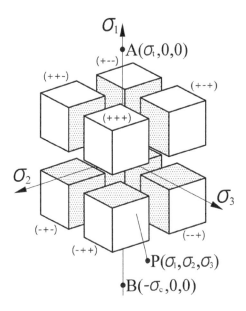

Figure 1.3 Eight quadrants in principal stress space

Furthermore, the cases of the combination of complex stresses (σ_1, σ_2, σ_3) are enormous.

Generally, the strength of materials under a complex stress state can be regarded as a function of stress state. This function is generally referred to as a strength theory. Sometimes the strength function is also referred to as yield criterion, failure criterion or strength hypothesis. It can be expressed by three principal stresses and several material parameters, as follows:

$$F\left(\sigma_1, \sigma_2, \sigma_3; K_1, K_2\right) = 0 \qquad (1.1)$$

The parameter of strength of materials should be obtained by common and well-established experiments. Generally, the tensile strength σ_t and compressive strength σ_c are easy to obtain, hence Eq. 1.1 can be expressed as a function of three principal stresses and two material parameters:

$$F\left(\sigma_1, \sigma_2, \sigma_3; \sigma_t, \sigma_c\right) = 0 \qquad (1.2)$$

If $\sigma_t = \sigma_c$, the two parameters can be simplified as one, i.e. $\sigma_t = \sigma_c = \sigma_y$, and the equation may be written as:

$$F(\sigma_1, \sigma_2, \sigma_3; \sigma_y) = 0 \tag{1.3}$$

1.3   Single strength theory

A lot of yield criteria and failure criteria have been proposed. Most of them can only be adopted for one kind of material. These criteria can be referred as the single strength theories or single failure criteria. Some basic strength theories were proposed, including maximum principal stress theory (Rankine, 1861), maximum principal strain theory (de Saint-Venant, 1870), maximum shear stress theory (Tresca, 1864; Guest, 1900), maximum distortion energy theory (Huber, 1904; Mises, 1913), maximum deviatoric stress theory (Burzynski, 1928; Schmidt, 1932; Ishlinsky, 1940; Hill, 1950; Haythornthwaite, 1961) or twin-shear yield criterion (Yu, 1961, 1983), Mohr-Coulomb criterion (Coulomb, 1773; Mohr, 1900), twin-shear strength theory (Yu *et al.*, 1985) and so on.

Maximum principal stress theory (Rankine criterion)

This assumes that the yield occurs when the largest principal stress exceeds the uniaxial tensile yield strength, that is,

$$\sigma_1 = \sigma_y \tag{1.4}$$

Maximum principal strain theory (Mariotte-St. Venant criterion)

This assumes that the yield occurs when the maximum principal strain reaches the strain corresponding to the yield point during a simple tensile test. It can be expressed in the principal stress state as

$$\sigma_1 - \nu(\sigma_2 + \sigma_3) = \sigma_y \tag{1.5}$$

Maximum shear stress theory (Tresca criterion)

This assumes that the yield occurs when the shear stress exceeds the shear yield strength. In principal stress state, it can be expressed as

$$\sigma_1 - \sigma_3 = \sigma_y \tag{1.6}$$

Maximum distortion energy theory (Huber-Mises criterion)

This assumes that the yield occurs when the distortion component exceeds that at the yield point for a simple tensile test. It can be expressed as

$$\frac{1}{\sqrt{2}}\left[(\sigma_1 - \sigma_2)^2 + (\sigma_2 - \sigma_3)^2 + (\sigma_3 - \sigma_1)^2\right]^{1/2} = \sigma_y \qquad (1.7)$$

Maximum deviatoric stress theory or twin-shear yield criterion

The maximum deviatoric stress theory assumes that the yield occurs when the maximum deviatoric stress reaches a critical value, that is,

$$\sigma_1 - \frac{1}{3}(\sigma_1 + \sigma_2 + \sigma_3) = C \qquad (1.8)$$

where C is a material parameter to be determined.

The twin-shear criterion assumes that yield occurs when the sum of two larger principal shear stresses reaches a critical value. It can be expressed in the principal stress state as

$$f = \sigma_1 - \frac{1}{2}(\sigma_2 + \sigma_3) = \sigma_y, \text{ when } \sigma_2 \leq \frac{1}{2}(\sigma_1 + \sigma_3) \qquad (1.9a)$$

$$f' = \frac{1}{2}(\sigma_1 + \sigma_2) - \sigma_3 = \sigma_y, \text{ when } \sigma_2 \geq \frac{1}{2}(\sigma_1 + \sigma_3) \qquad (1.9b)$$

These single strength theories (i.e. Rankine criterion, Mariotte-St. Venant criterion, Tresca criterion, Huber-Mises criterion and maximum deviatoric stress theory or twin-shear criterion) can be adopted for those materials having identical strength in tension and compression (also called non-SD material). Detailed descriptions of these strength theories can be found in many books (Zyczkowski, 1981; Yu, 1998). The limit loci of these single strength theories are illustrated in Figure 1.4.

Mohr-Coulomb strength theory

This assumes that failure occurs when the shear stress reaches a value that depends linearly on the normal stress. It can be expressed in the principal stress state as

$$\sigma_1 - \alpha\sigma_3 = \sigma_t \qquad (1.10)$$

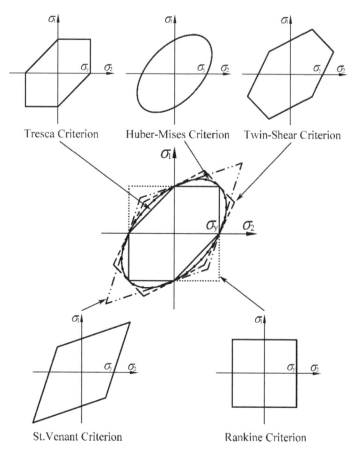

Figure 1.4 Limit loci of the single strength theories for non-SD material in the plane stress state

Twin-shear strength theory

This assumes that failure occurs when the combination of the two large principal shear stresses and the corresponding normal stresses reaches a critical value. It can be expressed in the principal stress state as

$$f = \sigma_1 - \frac{\alpha}{2}(\sigma_2 + \sigma_3) = \sigma_t, \text{ when } \sigma_2 \leq \frac{\sigma_1 + \alpha\sigma_3}{1+\alpha} \tag{1.11a}$$

$$f' = \frac{1}{2}(\sigma_1 + \sigma_3) - \alpha\sigma_3 = \sigma_t, \text{ when } \sigma_2 \geq \frac{\sigma_1 + \alpha\sigma_3}{1+\alpha} \tag{1.11b}$$

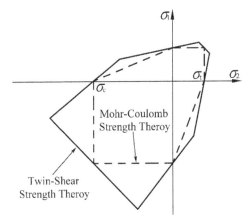

Figure 1.5 Limit loci of single strength theories for SD material in the plane stress state

Mohr-Coulomb strength theory and twin-shear theory can be adopted for materials having different strengths in uniaxial tension and compression (also called SD material). The limit loci in the plane stress state are shown in Figure 1.5.

1.4 Voigt-Timoshenko conundrum

A unified strength theory that can be used for more kinds of materials was considered impossible.

In 1901, Professor Voigt at Göttingen University, Germany, made a conclusion that it is impossible to formulate a single strength criterion that can be applied to various materials (Voigt, 1901).

Professor Timoshenko at Stanford University also said that

"A number of tests were made with combined stresses with a view to checking Mohr's theory. All these tests were made with brittle materials and the results obtained were not in agreement with theory. Voigt came to the conclusion that the question of strength is too complicated and that it is impossible to devise a single theory for successful application to all kinds of structural materials." (Timoshenko, 1953, History of Strength of Materials). This is the Voigt-Timoshenko conundrum.

In 1968, Mendelson said in his textbook *Plasticity* that "From an engineering viewpoint the accuracy of the von Mises criterion for yielding is amply sufficient, the search for more accurate theories, particularly since they are bound to be more complex, seems to be a rather thankless task" (Mendelson, 1968). This means that a simpler and better theory than the Mises criterion is impossible.

In 1984, Bowes, Russell and Suter stated in *Mechanics of Engineering Materials* that "An understanding of failure theory is necessary in order to avoid making some serious errors. Unfortunately, no single theory will be found to apply in all cases" (Bowes *et al.*, 1984, p. 247).

In 1985, the same thought was expressed in *Encyclopedia of China* (1985): "it is impossible to establish a unified strength theory for

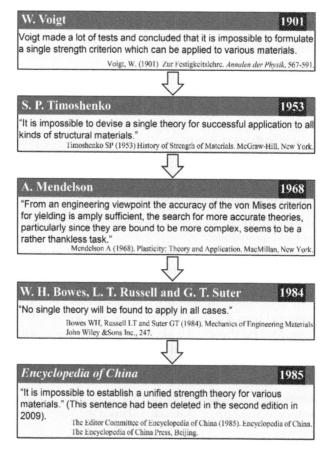

Figure 1.6 Voigt-Timoshenko conundrum

various materials." In the second edition of *Encyclopedia of China* (2009), this statement was removed.

A similar idea was expressed by Yu *et al.* (1985):

> because of the complexity and varieties of yields and failure of various materials and slip of crystals, it seems impossible to expound these phenomena by a single criterion.

This is a conundrum that related to strength theory in the 20th century, and it was unsolved until 1990. It is briefly summarized in Figure 1.6.

1.5 The development of unified strength theory

To find a solution to the unified strength theory that is suitable for different kinds of materials, we need to understand the common behaviors of the strength criteria. The Drucker postulate, which was found in 1951, provides a theoretical frame relating to the yield criterion. The convexity of yield surface for material under complex stress appeared as well. The study of yield criterion may be developing on a more reliable theoretical basis.

It is difficult and interesting to establish a unified strength theory, which is expected to have the following characteristics:

1   It should conform to the Drucker postulate. The yield loci must be convex and cover all regions from the lower bound to the upper bound.
2   It should be physically meaningful and have a unified mechanical model; it is better to express the strength theory by a simple unified mathematical expression.
3   It should agree with the experimental results of various materials under the complex stress state.
4   It should take into account the influence of each stress components σ_1, σ_2 and σ_3 on the material failure.
5   It should be convenient for application to analytical solutions and numerical solutions.

Unified strength theory proposed by Yu and He (1991) is a set of yield criteria and failure criteria that can be adopted for most materials. The expression of unified strength theory is shown in Eqs. 1.12a and 1.12b. It can be seen that the mathematical expressions of twin-shear yield criterion (Yu, 1961, 1983), twin-shear strength theory

(Yu *et al.*, 1985) and unified strength theory are very similar. The differences among these expressions reflect the development history of unified strength theory.

$$F = \sigma_1 - \frac{\alpha}{1+b}(b\sigma_2 + \sigma_3) = \sigma_t, \text{ when } \sigma_2 \leq \frac{\sigma_1 + \alpha\sigma_3}{1+\alpha} \qquad (1.12a)$$

$$F' = \frac{1}{1+b}(\sigma_1 + b\sigma_2) - \alpha\sigma_3 = \sigma_t, \text{ when } \sigma_2 \geq \frac{\sigma_1 + \alpha\sigma_3}{1+\alpha} \qquad (1.12b)$$

The unified strength theory is also referred to as Yu-unified strength theory (Yu-UST).

1.6   Applications of the Yu-UST

The unified strength theory has been widely applied in many fields. Hundreds of papers have been published in different journals and conference proceedings (Altenbach and Ochsner, 2014; Kolupaev and Altenbach, 2009, 2010; Kolupaev *et al.*, 2013; Ma *et al.*, 2014). The unified strength theory is also used in master's theses and doctoral dissertations. The applied fields include mechanics of materials, soil mechanics, rock mechanics, concrete mechanics, plasticity, geotechnical plasticity, civil engineering, mechanical engineering, aerospace engineering and so on. Many applications of the Yu-unified strength theory can be seen in books such as *Generalized Plasticity* (Yu, 2006), *Structural Plasticity* (Yu *et al.*, 2009), *Computational Plasticity: with Emphasis on the Applications of the Unified Strength Theory* (Yu and Li, 2012), *Numerical Analysis of Geo-materials Under Complex Stress* (Ma *et al.*, 2017) and *Strength Theory of Geotechnical Structure* (Fan *et al.*, 2018).

The applications of Yu-UST are summarized in Figure 1.7. Several applications are described in Chapter 5.

1.7   Summary

It has been a long process for the development of single strength theories and unified strength theory (Marin, 1936; Paul, 1968a, 1968b; Pisarenko and Lebedev, 1969; Yu, 1961, 1983, 2018; Yu *et al.*, 1985; Yu and He, 1991). Yu-UST is a set of yield criteria and failure criteria of materials under complex stress state. It can be widely used for various materials in solid mechanics and engineering. The Yu-UST is the solution to the Voigt-Timoshenko conundrum.

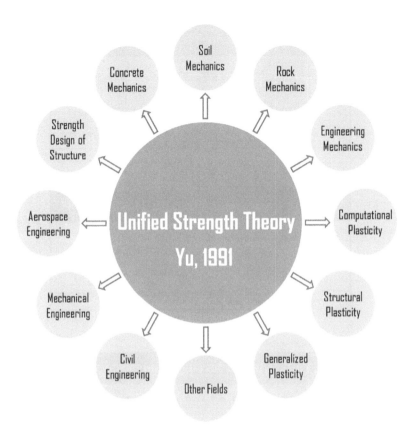

Figure 1.7 Applications of the Yu-UST

A monograph related to the Yu-UST was published in Chinese in 1992 (Yu, 1992). The title of the monograph is *New System of Strength Theory*. Another monograph, *Unified Strength Theory and Its Applications*, was published by Springer in 2004 (Yu, 2004). These two books won the National Natural Science Prize of China in 2011. The National Natural Science prize is the highest academic achievement prize in china.

The development of Yu-UST is summarized in Figure 1.8. It is seen that these three formulae are very similar. Only two variables, α and b, have been added to the original formula over the past 30 years. How slow it is!

M-H Yu   Twin-Shear Yield Criterion   1961

$$f = \sigma_1 - \frac{1}{2}(\sigma_2 + \sigma_3) = \sigma_y, \text{ when } \sigma_2 \leq \frac{1}{2}(\sigma_1 + \sigma_3)$$

$$f' = \frac{1}{2}(\sigma_1 + \sigma_2) - \sigma_3 = \sigma_y, \text{ when } \sigma_2 \geq \frac{1}{2}(\sigma_1 + \sigma_3)$$

Yu MH (1961) Plastic potential and flow rules associated singular yield criterion. *Res. Report of Xi'an Jiaotong University*. Xi'an, China (in Chinese)

M-H Yu   Twin-Shear Strength Theory   1985

$$f = \sigma_1 - \frac{\alpha}{2}(\sigma_2 + \sigma_3) = \sigma_t, \text{ when } \sigma_2 \leq \frac{\sigma_1 + \alpha\sigma_3}{1+\alpha}$$

$$f' = \frac{1}{2}(\sigma_1 + \sigma_2) - \alpha\sigma_3 = \sigma_t, \text{ when } \sigma_2 \geq \frac{\sigma_1 + \alpha\sigma_3}{1+\alpha}$$

Yu MH, He LN, Song LY (1985) Twin shear stress theory and its generalization. *Science in China Series A*, 28(11), 1174-1183.

M-H Yu   Unified Strength Theory   1991

$$F = \sigma_1 - \frac{\alpha}{1+b}(b\sigma_2 + \sigma_3) = \sigma_t, \text{ when } \sigma_2 \leq \frac{\sigma_1 + \alpha\sigma_3}{1+\alpha}$$

$$F' = \frac{1}{1+b}(\sigma_1 + b\sigma_2) - \alpha\sigma_3 = \sigma_t, \text{ when } \sigma_2 \geq \frac{\sigma_1 + \alpha\sigma_3}{1+\alpha}$$

Yu MH, He LN (1991) A new model and theory on yield and failure of materials under the complex stress state. *Mechanical Behavior of Materials-6* (ICM-6). Jono M and Inoue T eds. Pergamon Press, Oxford, (3):841–846.

Figure 1.8 The development of Yu-UST

References

Altenbach, H. & Ochsner, A. (2014) *Plasticity of Pressure-Sensitive Materials*. Springer, Berlin.

Bowes, W.H., Russell, L.T. & Suter, G.T. (1984) *Mechanics of Engineering Materials*. John Wiley & Sons Inc., Hoboken, NJ, p. 247.

Burzynski, W.T. (1928) *Studium nad hipotezami wytezenia* (Study on Material Effort Hypotheses), Doctoral Dissertation Printed in Polish by the Academy of Technical Sciences Lwów, 1928, 1–192. The Burzynski's doctoral dissertation was published recently in English in 2009 in Poland (Translated by Teresa Fras and Anna Strek, Engineering Transactions, 57, 127–157).

Coulomb, C.A. (1773, 1776) Essai Sur une application des regles de maximis and minimis a quelques problemes de statique, relatifs a l'a rchitecture. Memoires de Mathematique and de Physique, presentes a l' Academie. *Royale des Sciences par divers Savans, and lus dans ses Assemblees*, 7, 343–382, Paris (1776) (English translation: Note on an application of the rules of maximum and minimum to some statical problems, relevant to architectuture, see Heyman, J. (1977) *Coulomb'ss Menoir on Statics*. Imprerial College Press, London, pp. 41–74).

de Saint-Venant, B. (1870) Memoire sur l'establissement des equations differentielles des mouvements interieurs operes dans les corps solides ductiles au dela des limites ou l' elasticite pourrait les ramener a leur premier etat. *Comptes Rendus hebdomadaires des Seances de l'Academie des Sciences*, 70, 473–480.

The Editor Committee of Encyclopedia of China (1985) *Encyclopedia of China*. The Encyclopedia of China Press, Beijing.

The Editor Committee of Encyclopedia of China (2009) *The Second Edition of Encyclopedia of China*. The Encyclopedia of China Press, Beijing.

Fan, W., Yu, M.H. & Deng, L.S. (2018) *Strength Theory of Geotechnical Structure*. Science Press, Beijing.

Guest, J.J. (1900) On the strength of ductile materials under combined stress. *Phil. Mag. and J. Sci.*, 69–133.

Haythornthwaite, R.M. (1961) Range of yield condition in ideal plasticity. *J. Engrg. Mech.* ASCE, 87(6), 117–133.

Hill, R. (1950) *Philosophisical Magazine (London)*, 41, 733–744 (See: Zyczkowski M 1981, p. 100).

Huber, M.T. (1904) Przyczynek do podstaw wytorymalosci. *Czasopismo Technizne*, 22(81), (Lwow, 1904); Pisma, 2, PWN, Warsaw, 1956.

Ishlinsky, A.Y. (1940) Hypothesis of strength of shape change. *Uchebnye Zapiski Moskovskogo Universiteta, Mekhanika*, 46 (in Russian, from Zyczkowski, 1981, p. 100: and Pisarenko and Lebedev 1976, Chinese edition 1983, pp. 70–71).

Kolupaev, V.A. & Altenbach, H. (2009) Application of the unified strength theory of Mao-Hong Yu to plastics. In: Grellmann, W. (ed) *Tagung Deformations- und Bruchverhalten von Kunststoffen 24–26.06.2009*, Book of Abstracts. Martin-Luther-Universität Halle-Wittenberg, Merseburg, pp. 320–339 (in German).

Kolupaev, V.A. & Altenbach, H. (2010) Einige Überlegungen zur Unified Strength Theory von Mao-Hong Yu (Considerations on the unified strength theory due to Mao-Hong Yu). *Forschung im Ingenieurwesen (Forsch Ingenieurwes)*, 74(Springer), 135–166 (in German, English Abstract).

Kolupaev, V.A., Yu, M.H. & Altenbach, H. (2013) Visualization of the unified strength theory. *Archive of Applied Mechanics*, 83, 1061–1085.

Ma, Z.Y., Liao, H.J. & Dang, F.N. (2014) Influence of intermediate principal stress on the bearing capacity of strip and circular footings. *J. of Engineering Mechanics*, ASCE, 140(7), 04014041, 1–14.

Ma, Z.Y., Liao, H.J. & Qi, J.L. (2017) *Numerical Analysis of Geo-Materials under Complex Stress.* Xi'an Jiaotong University Press, Xi'an.

Marin, J. (1936) Failure theories of materials subjected to combined stresses. *Trans. Am. Soc. Civ. Engrs.* 101, 1162–1194.

Mendelson, A. (1968) *Plasticity: Theory and Application.* MacMillan, New York, p. 62.

Mohr, O. (1900) Welche Umstande bedingen die Elastizitatsgrenze und den Bruch eines Materials? *Zeitschrift des Vereins deutscher Ingenieure,* 44, 1524–1530, 1572–1577.

Paul, B. (1968a) Generalized pyramidal fracture and yield criteria. *Int. J. of Solids and Structures,* 4, 175–196.

Paul, B. (1968b) Macroscopic criteria for plastic flow and brittle fracture. In: Liebowitz, H. (ed) *Fracture, an Advanced Treatise,* Vol.2. Academic, New York, pp. 313–496.

Pisarenko, G.S. & Lebedev, A.A. (1969) *Deformation and Fracture of Materials under Combined Stress.* Izd. Naukoea Dumka, Kiev (in Russian).

Rankine, W.J.M. (1861) *Manual of Applied Mechanics,* 21st edition), 1921.

Schmidt, R. (1932) Über den Zusammenhang von Spannungen und Formaenderungen im Verfestigungsgebiet, IA 3: 215–235 (in German, from Zyczkowski, 1981, p. 100).

Timoshenko, S.P. (1953) *History of Strength of Materials.* McGraw-Hill, New York.

Tresca, H. (1864) Sur l'e coulement des corps solids soumis a de fortes pression. *Comptes Rendus hebdomadaires des Seances de l' Academie des Sciences,* 59, 754–758.

Voigt, W. (1901) Zur Festigkeitslehre. *Annalen der Physik,* 567–591.

von Mises, R. (1913) Mechanik der festen Körper im plastisch deformablen Zustand. Nachrichten von der Königlichen Gesellschaft der wissenschaften zu Göettinger. *Mathematisch-Physikalische Klasse,* 582–592.

Yu, M.H. (1961) *General Behavior of Isotropic Yield Function.* Res. Report of Xi'an Jiaotong University, Xi'an, China (in Chinese).

Yu, M.H. (1983) Twin shear stress yield criterion. *Int. J. of Mech. Sci.* 25, 71–74.

Yu, M.H. (1992) *A New System of Strength Theory.* Xian Jiaotong University Press, Xian, China (in Chinese).

Yu, M.H. (1998) *Twin-Shear Strength Theory and Its Applications.* Science Press, Beijing (in Chinese).

Yu, M.H. (2004) *Unified Strength Theory and Its Applications.* Springer, Berlin.

Yu, M.H. (2006) *Generalized Plasticity.* Springer, Berlin.

Yu, M.H. (2018) *Unified Strength Theory and Its Applications,* 2nd edition. Springer and Xi'an Jiaotong University Press, Singapore and Xi'an.

Yu, M.H. & He, L.N. (1991) A new model and theory on yield and failure of materials under the complex stress state. In: Jono, M. and Inoue, T. (eds) *Mechanical Behavior of Materials-6 (ICM-6),* Vol. 3. Pergamon Press, Oxford, pp. 841–846.

Yu, M.H. & Li, J.C. (2012) *Computational Plasticity: With Emphasis on the Applications of the Unified Strength Theory*. Springer and Zhejiang University Press, Berlin and Hangzhou.

Yu, M.H., He, L.N. & Song, L.Y. (1985) Twin shear stress theory and its generalization. *Science in China Series A, English Edition*, 28(11), 1174–1183.

Yu, M.H., Ma, G.W. & Li, J.C. (2009) *Structural Plasticity*. Springer and Zhejiang University Press, Berlin and Hangzhou.

Zyczkowski, M. (1981) *Combined Loadings in the Theory of Plasticity*. Polish Scientific Publishers, PWN and Nijhoff.

Readings

[**Readings 1–1**] "The value of b ranges from 0 to 1 for any convex strength envelope. When the effect of the intermediate principal shear stress is ignored ($b = 0$), the UTSS expression is equivalent to Mohr-Coulomb's criterion. When the effect of the intermediate principal shear stress grows to its fullest ($b = 1$), it is Yu's primitive twin-shear strength (TSS) criterion. For all materials, the effect of the intermediate principal shear stress varies between 0 and 100%; that is, $0 \le b \le 1$. Different values of the weighting coefficient b generate a series of new strength criteria. Any existing criterion may find a particular value of b to match its shape function in the deviatoric plane."

"Yu's UTSS theory has a sound theoretical background."

Fan, S.C. & Wang, F. (2002) A new strength criterion for concrete. *ACI Structural J.*, 99, 317–326.

2 Stress state and elements

2.1 Introduction

The strength of material is related to the stresses acting on the object (Bowes *et al.*, 1984; Means, 1976; Timoshenko, 1972). A simple example of the stress state can be illustrated by a bar subjected to the uniaxial tension as shown in Figure 2.1.

There is only a normal stress acting on the transverse section, m-m, of the bar. However, normal stress σ_α and shear stress τ_α are acting on the oblique section as well. On the same point, the magnitude and

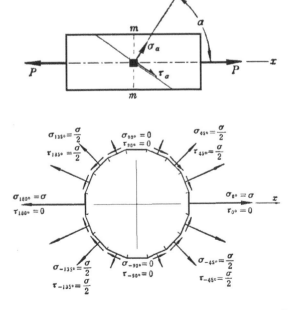

Figure 2.1 The stress state of the element subjected to the uniaxial tension

direction of the stress acting on the cross section changed with the angle α, as shown in Figure 2.1. The maximum normal stress acts on the transverse section ($\sigma_{max} = \sigma = F/A$, $\tau = 0$); the maximum shear stress acts on the cross section when $\alpha = 45°$ ($\tau_{max} = \tau_{45°} = \sigma/2$). The normal stress is a function of α as shown in Figure 2.2a. The shear stress is a function of α as shown in Figure 2.2b.

The section that only has the normal stress is called the principal plane, and the stress acting on the principal plane is called the principal stress. The most general state of stress is the three-dimensional principal stresses (σ_1, σ_2, σ_3), as shown in Figure 2.3.

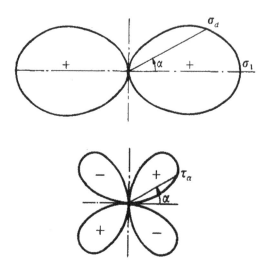

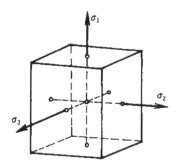

Figure 2.2 The change of normal stress and shear stress with α at a point

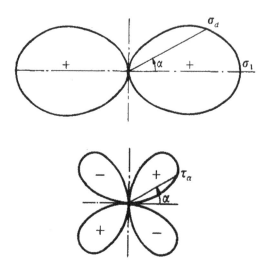

Figure 2.3 Principal stress element

2.2 Stresses on the oblique plane

If the three principal stresses σ_1, σ_2 and σ_3 acting on three principal planes, respectively, at a point are given, we can determine the stresses acting on any plane through this point. This can be done by consideration of the static equilibrium of a tetrahedron formed by this plane and the principal planes, as shown in Figure 2.4. In this figure, we have shown the principal stresses acting on the three principal planes. These stresses are assumed to be known. We wish to find the stresses σ_α, τ_α acting on the oblique plane whose normal has direction cosines l, m and n (Figure 2.4).

The normal stress σ_α and shear stress τ_α acting on this plane can be determined as follows:

$$\sigma_\alpha = \sigma_1 l^2 + \sigma_2 m^2 + \sigma_3 n^2 \tag{2.1}$$
$$\tau_\alpha = \sigma_1{}^2 l^2 + \sigma_2{}^2 m^2 + \sigma_3{}^2 n^2 - (\sigma_1 l^2 + \sigma_2 m^2 + \sigma_3 n^2) \tag{2.2}$$
$$\vec{p}_\alpha = \vec{\sigma}_\alpha + \vec{\tau}_\alpha \tag{2.3}$$

The three principal shear stresses τ_{13}, τ_{12} and τ_{23} can be obtained as follows:

$$\tau_{13} = \frac{1}{2}(\sigma_1 - \sigma_3); \; \tau_{12} = \frac{1}{2}(\sigma_1 - \sigma_2); \; \tau_{23} = \frac{1}{2}(\sigma_2 - \sigma_3) \tag{2.4}$$

The maximum shear stress acts on the plane bisecting the angle between the largest and smallest principal stresses is equal to half of the difference between these principal stresses, that is,

$$\tau_{\max} = \tau_{13} = \frac{1}{2}(\sigma_1 - \sigma_3) \tag{2.5}$$

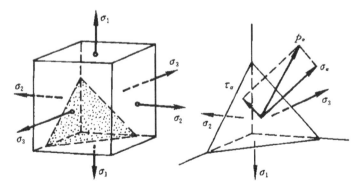

Figure 2.4 Stress on an infinitesimal tetrahedron

Table 2.1 Direction cosines of the principal planes and the principal shear stress planes

	Principal plane			Principal shear stress plane		
$l =$	± 1	0	0	$\pm\dfrac{1}{\sqrt{2}}$	$\pm\dfrac{1}{\sqrt{2}}$	0
$m =$	0	± 1	0	$\pm\dfrac{1}{\sqrt{2}}$	0	$\pm\dfrac{1}{\sqrt{2}}$
$n =$	0	0	± 1	0	$\pm\dfrac{1}{\sqrt{2}}$	$\pm\dfrac{1}{\sqrt{2}}$
$\sigma =$	σ_1	σ_2	σ_3	$\sigma_{12} = \dfrac{\sigma_1+\sigma_2}{2}$	$\sigma_{13} = \dfrac{\sigma_1+\sigma_3}{2}$	$\sigma_{23} = \dfrac{\sigma_2+\sigma_3}{2}$
$\tau =$	0	0	0	$\tau_{12} = \dfrac{\sigma_1-\sigma_2}{2}$	$\tau_{13} = \dfrac{\sigma_1-\sigma_3}{2}$	$\tau_{23} = \dfrac{\sigma_2-\sigma_3}{2}$

The corresponding normal stresses σ_{13}, σ_{12} and σ_{23} acting on the sections, where τ_{13}, τ_{12} and τ_{23} are acting, respectively, are

$$\sigma_{13} = \frac{1}{2}(\sigma_1 + \sigma_3); \; \sigma_{12} = \frac{1}{2}(\sigma_1 + \sigma_2); \; \sigma_{23} = \frac{1}{2}(\sigma_2 + \sigma_3) \qquad (2.6)$$

It is seen from Eq. 2.4 that the maximum principal shear stress τ_{13} is equal to the sum of the other two $(\tau_{12} + \tau_{23})$, i.e.

$$\tau_{13} = \tau_{12} + \tau_{23} \qquad (2.7)$$

This means that there are only two independent variables among these three principal shear stresses.

The direction cosines l, m and n of the principal plane, principal shear stress plane and the octahedral plane, as well as the principal shear stresses and corresponding normal stresses are listed in Table 2.1.

2.3   Cubic hexahedron and octahedron

According to the stress state, various polyhedral elements can be illustrated, as shown in Figures 2.5, 2.6 and 2.7.

The principal stress element is a cubic element, and the three principal stresses σ_1, σ_2 and σ_3 act on this element. The principal stress element and three principal stresses $(\sigma_1, \sigma_2, \sigma_3)$ are shown in Figure 2.3.

The maximum shear stress element $(\tau_{13}, \sigma_{13}, \sigma_2)$ is a quadrangular prism element, which the maximum shear stress τ_{13}, corresponding normal stress σ_{13}, as well as the intermediate principal stress σ_2 act on. This kind of element, as shown in Figure 2.5a, may be referred to as the single-shear element because only one shear stress and corresponding normal stress act on the element.

The quadrangular prism element $(\tau_{12}, \sigma_{12}, \sigma_3)$ is shown in Figure 2.5b, including the intermediate principal shear stress τ_{12} and the corresponding normal stress σ_{12}, as well as the minimum principal stress σ_3. The quadrangular prism element $(\tau_{23}, \sigma_{23}, \sigma_1)$ is shown in Figure 2.5c, acted by the minimum principal shear stress τ_{23} and the corresponding normal stress σ_{23}, as well as the maximum principal stress σ_1.

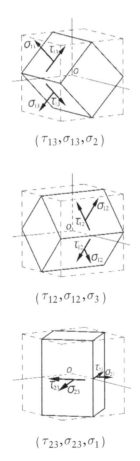

$(\tau_{13}, \sigma_{13}, \sigma_2)$

$(\tau_{12}, \sigma_{12}, \sigma_3)$

$(\tau_{23}, \sigma_{23}, \sigma_1)$

Figure 2.5 Single-shear elements

Figure 2.6a shows an orthogonal octahedron $(\tau_{13}, \tau_{12}; \sigma_{13}, \sigma_{12})$, and the principal shear stresses τ_{13}, τ_{12} and the corresponding normal stresses σ_{13}, σ_{12} act on this element. This new element was proposed by Yu (Yu *et al.*, 1985; Yu, 1998, 2004, 2007, 2011, 2018). It can be referred as the twin-shear element.

The principal shear stresses τ_{13}, τ_{23} and the corresponding normal stresses σ_{13}, σ_{23} act on an orthogonal octahedron element $(\tau_{13}, \tau_{23}; \sigma_{13}, \sigma_{23})$, as shown in Figure 2.6b. This element can also be referred to as the twin-shear element (Yu *et al.*, 1985). They are available to use for the mechanical model of strength theory.

The process of getting the twin-shear element from principal stress element to single-shear element and then from single-shear element to

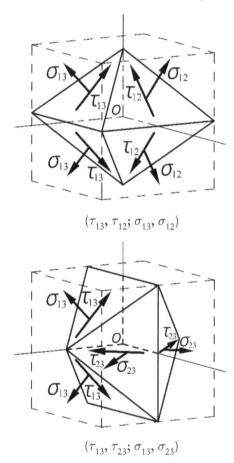

$(\tau_{13}, \tau_{12}; \sigma_{13}, \sigma_{12})$

$(\tau_{13}, \tau_{23}; \sigma_{13}, \sigma_{23})$

Figure 2.6 Twin-shear elements

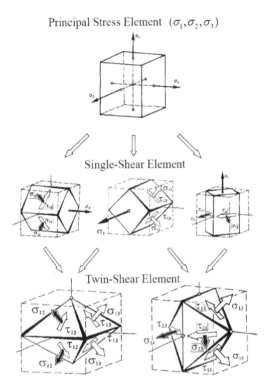

Principal Stress Element $(\sigma_1,\sigma_2,\sigma_3)$

Single-Shear Element

Twin-Shear Element

Figure 2.7 The process of getting the twin-shear element

twin-shear element is illustrated in Figure 2.7. Figure 2.7 also shows the relations among the cubic element, single-shear element and twin-shear element.

2.4 Mohr stress circle

The three principal stresses, three principal shear stresses and the three normal stresses acting on the principal shear stresses plane can be illustrated by three stress circles, which are referred to as the Mohr stress circles, as shown in Figure 2.8. The magnitude of the normal and shear stresses in any plane are equal to the distance of the corresponding stress point on the stress circle to the origin of the referred coordinate frame. The three principal shear stresses are equal to the radius of the three stress circles. Detailed descriptions of the stress circle can

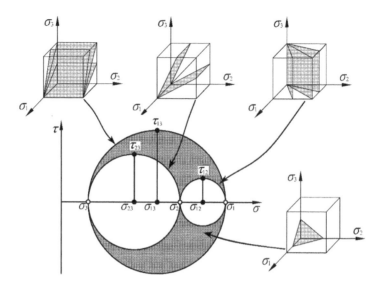

Figure 2.8 The concept of Mohr's stress circle

be found in books written by Johnson and Mellor (1962), Kussmaul (1981) and Chakrabarty (1987) and other textbooks of plasticity.

2.5   Stress space

The stress point $P\,(\sigma_1, \sigma_2, \sigma_3)$ in stress space can be expressed by other forms, such as $P(x, y, z)$ or $P(n, r, \theta)$. The geometrical representation of these transfers can be seen in Figure 2.9 and 2.10.

For the straight line OZ passing through the origin and making the same angle with each of the coordinate axes, the equation is

$$\sigma_1 = \sigma_2 = \sigma_3 \tag{2.9}$$

The equation of the π_0-plane is

$$\sigma_1 + \sigma_2 + \sigma_3 = 0 \tag{2.10}$$

The stress vector $\overrightarrow{OP}$ can be separated into two parts: the hydrostatic stress vector $\overrightarrow{ON}$ and the mean shear stress vector $\overrightarrow{NP}$.

$$\overrightarrow{OP} = \overrightarrow{ON} + \overrightarrow{NP} \tag{2.11}$$

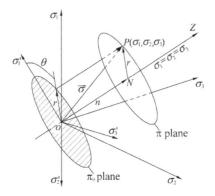

Figure 2.9 Cylindrical coordinates

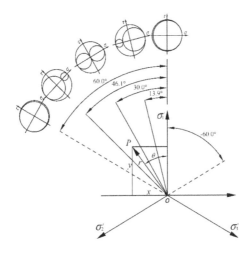

Figure 2.10 Stress state in the π-plane

The π-plane is parallel to the π_0-plane and is given by

$$\sigma_1 + \sigma_2 + \sigma_3 = C \tag{2.15}$$

in which C is a constant. The mean stress σ_m is the same for all points in the π-plane of stress space and

$$\sigma_m = \frac{1}{3}(\sigma_1 + \sigma_2 + \sigma_3) \tag{2.16}$$

The projections of the three principal stress axes in stress space σ_1, σ_2, σ_3 are σ_1', σ_2', σ_3'. The relationships among them are

$$\sigma_1' = \sigma_1 \cos\beta = \sqrt{\frac{2}{3}}\sigma_1$$

$$\sigma_2' = \sigma_2 \cos\beta = \sqrt{\frac{2}{3}}\sigma_2 \qquad (2.17)$$

$$\sigma_3' = \sigma_3 \cos\beta = \sqrt{\frac{2}{3}}\sigma_3$$

where β is the angle between $O'A$, $O'B$, $O'C$, as shown in Figure 2.11.
    The relations between the coordinates of the deviatoric plane (x,y,z) and the principal stresses $(\sigma_1, \sigma_2, \sigma_3)$ are

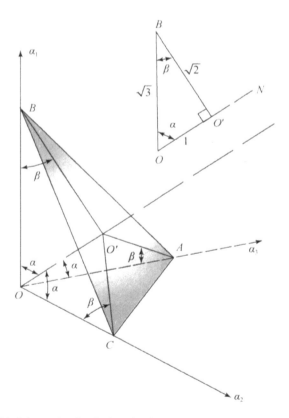

Figure 2.11 Schematic of a deviatoric plane

$$x = \frac{1}{\sqrt{2}}(\sigma_3 - \sigma_2)$$

$$y = \frac{1}{\sqrt{6}}(2\sigma_1 - \sigma_2 - \sigma_3) \qquad (2.18)$$

$$z = \frac{1}{\sqrt{3}}(\sigma_1 + \sigma_2 + \sigma_3)$$

or

$$\sigma_1 = \frac{1}{3}\left(\sqrt{6}y + \sqrt{3}z\right)$$

$$\sigma_2 = \frac{1}{6}\left(2\sqrt{3}z - \sqrt{6}y - 3\sqrt{2}x\right) \qquad (2.19)$$

$$\sigma_3 = \frac{1}{6}\left(3\sqrt{2}x - \sqrt{6}y + 2\sqrt{3}z\right)$$

The relations between the cylindrical coordinates (n, r, θ) and the principal stresses ($\sigma_1, \sigma_2, \sigma_3$) are

$$n = \left|\overrightarrow{ON}\right| = \frac{1}{\sqrt{3}}(\sigma_1 + \sigma_2 + \sigma_3) \qquad (2.20)$$

$$r = \left|\overrightarrow{NP}\right| = \frac{1}{\sqrt{3}}\left[(\sigma_1 - \sigma_2)^2 + (\sigma_2 - \sigma_3)^2 + (\sigma_3 - \sigma_1)^2\right]^{\frac{1}{2}} \qquad (2.21)$$

$$\theta = \arccos\left(\frac{2\sigma_1 - \sigma_2 - \sigma_3}{\sqrt{2}\sqrt{(\sigma_1 - \sigma_2)^2 + (\sigma_2 - \sigma_3)^2 + (\sigma_3 - \sigma_1)^2}}\right) \qquad (2.22)$$

These relations are suitable to the conditions $\sigma_1 \geq \sigma_2 \geq \sigma_3$ and $0 \leq \theta \leq \pi/3$. The limit loci in the π-plane has threefold symmetry, so if the limit loci in the range of 60° are given, then the limit loci in π-plane can be obtained.

The three principal stresses can be expressed by the cylindrical coordinate as follows:

$$\sigma_1 = \frac{1}{\sqrt{3}}n + \sqrt{\frac{2}{3}}r\cos\theta$$

$$\sigma_2 = \frac{1}{\sqrt{3}}n + \sqrt{\frac{2}{3}}r\cos(\theta - 2\pi/3), \ 0 \leq \theta \leq \frac{\pi}{3} \qquad (2.23)$$

$$\sigma_3 = \frac{1}{\sqrt{3}}n + \sqrt{\frac{2}{3}}r\cos(\theta + 2\pi/3)$$

2.6   Summary

The concept of stress state is very important for understanding the concept of strength theory. Stress state has been studied in mechanics of materials. There are, however, some differences here. The single-shear element, twin-shear element and stress space are described in this chapter.

The idea of the twin-shear element is very important, and it can be found in an article by Yu *et al.* (1985). It will be used to introduce a new strength theory.

References

Bowes, W.H., Russell, L.T. & Suter, G.T. (1984) *Mechanics of Engineering Materials*. John Wiley & Sons Inc., Hoboken, NJ, p. 247.

Chakrabarty, J. (1987) *Theory of Plasticity*. McGraw-Hill, New York.

Johnson, W. & Mellor, P.B. (1962) *Plasticity for Mechanical Engineers*. Van Nostrand, London and New York.

Kussmaul, K. (1981) *Festigkeitslehre I*. MPA Stuttgart, University Stuttgart.

Means, W.D. (1976) *Stress and Strain*. Springer-Verlag, Inc., New York.

Timoshenko, S. & Gere, J. (1972) *Mechanics of Materials*. D. Van Nostrand Reinhold Company Ltd.

Yu, M.H. (1998) *Twin-Shear Strength Theory and Its Applications*. Science Press, Beijing (in Chinese).

Yu, M.H. (2004) *Unified Strength Theory and Its Applications*. Springer, Berlin.

Yu, M.H. (2007) Linear and nonlinear unified strength theory. *Chinese Journal of Rock Mechanics and Engineering*, 26(4), 662–669. (in chinese).

Yu, M.H. (2011) *New System of Strength Theory: Theory, Development and Applications*. Xian Jiaotong University Press, Xian, China (in Chinese).

Yu, M.H. (2018) *Unified Strength Theory and Its Applications*, 2nd edition. Springer and Xi'an Jiaotong University Press, Singapore and Xi'an.

Yu, M.H., He, L.N. & Song, L.Y. (1985) Twin shear stress theory and its generalization. *Science in China Series A*, 28(11), 1174–1183.

Readings

[Readings 2–1] There are three pictures of the statue of Venus collected at Italian Naples Country Archaeology Museum, as shown in Figure 2.17. They are taken from the statue of Goddess of Venus by

Shu-Qi Yu from three directions. These images are different, but they are all the identical Venus at Naples. The concept of stress state is somewhat similar in this point.

Figure 2.12 The Love Goddess of Venus collected at Italian Naples Country Archaeology Museum (Shu-qi Yu took pictures from the three directions for Venus)

3 Yu-unified strength theory (Yu-UST)

3.1 Introduction

A lot of strength theories were proposed in the past. Most of them can be adopted for a single kind of material. The Tresca yield criterion can be suitable for non-SD materials with shear yield strength τ_y equaling the half of uniaxial yield strength of materials σ_y, i.e. $\tau_y = 0.5\sigma_y$. The Huber-Mises yield criterion can be adopted for non-SD materials with shear yield strength τ_y equaling 0.577 uniaxial yield strength of materials σ_y, i.e. $\tau_y = 0.577\sigma_y$. The twin-shear yield criterion can be adopted for non-SD materials with shear yield strength τ_y equaling 0.667 uniaxial yield strength of materials σ_y, i.e. $\tau_y = 0.667\sigma_y$. The Mohr-Coulomb strength theory is satisfied for the SD materials with $\tau_0 = \dfrac{\sigma_t \sigma_c}{\sigma_t + \sigma_c}$.

The development of the Yu-UST has been briefly described in Chapter 1. The characteristics, limit surfaces and loci, and verifications with experimental results of Yu-UST are described in detail in this chapter.

3.2 The Yu-unified strength theory

The stress state of an element has been described in the previous chapter. The principal stress state $(\sigma_1, \sigma_2, \sigma_3)$ can be converted into the principal shear stress state $(\tau_{13}, \tau_{12}, \tau_{23})$. The three shear stresses, however, only have two independent variables. The maximum principal shear stress τ_{13} equals the sum of the other two $(\tau_{12} + \tau_{23})$, i.e. $\tau_{13} = \tau_{12} + \tau_{23}$. Therefore, a new idea about twin shear is presented. The shear stress state can be converted into the two twin-shear stress states $(\tau_{13}, \tau_{12}; \sigma_{13}, \sigma_{12})$ and $(\tau_{13}, \tau_{23}; \sigma_{13}, \sigma_{23})$. It can be illustrated by the two new orthogonal octahedral elements as shown in Figures 3.1(a) and 3.1(b). The process of getting the twin-shear element can be seen in Figure 2.7 in the previous chapter.

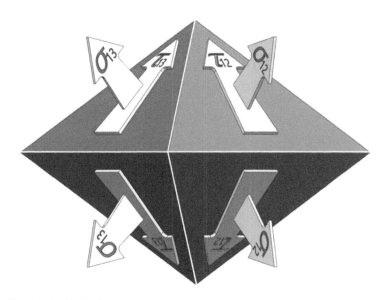

Figure 3.1a Twin-shear element (τ_{13}, τ_{12}; σ_{13}, σ_{12})

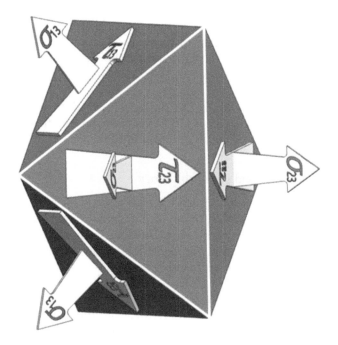

Figure 3.1b Twin-shear element (τ_{13}, τ_{23}; σ_{13}, σ_{23})

Considering the all stress components acting on the twin-shear elements and the different effects of various stresses on the failure of materials, the Yu-UST can be established as follows:

$$F = \tau_{13} + b\tau_{12} + \beta(\sigma_{13} + b\sigma_{12}) = C, \text{ when}$$
$$\tau_{12} + \beta\sigma_{12} \geq \tau_{23} + \beta\sigma_{23} \tag{3.1a}$$

$$F' = \tau_{13} + b\tau_{23} + \beta(\sigma_{13} + b\sigma_{23}) = C, \text{ when}$$
$$\tau_{12} + \beta\sigma_{12} \leq \tau_{23} + \beta\sigma_{23} \tag{3.1b}$$

in which β is the coefficient that represents the effect of the normal stress on failure, C is a strength parameter of material and b is a parameter that reflects the influence of the intermediate principal shear stress τ_{12} or τ_{23} and the corresponding normal stress σ_{12} or σ_{23} on the failure of material. τ_{13}, τ_{12} and τ_{23} are principal shear stresses, and σ_{13}, σ_{12} and σ_{23} are the corresponding normal stresses acting on the sections where τ_{13}, τ_{12} and τ_{23} are acting. It is the basic idea of the Yu-UST (Yu and He, 1991; Yu, 1992).

The relationships among the stresses acting on the two twin-shear elements and principal stresses are

$$\tau_{13} = \frac{1}{2}(\sigma_1 - \sigma_3), \ \sigma_{13} = \frac{1}{2}(\sigma_1 + \sigma_3) \tag{3.2a}$$

$$\tau_{12} = \frac{1}{2}(\sigma_1 - \sigma_2), \ \sigma_{12} = \frac{1}{2}(\sigma_1 + \sigma_2) \tag{3.2b}$$

$$\tau_{23} = \frac{1}{2}(\sigma_2 - \sigma_3), \ \sigma_{23} = \frac{1}{2}(\sigma_2 + \sigma_3) \tag{3.2c}$$

The parameters of β and C in Eqs. 3.1a and 3.1b can be determined by experimental results of uniaxial tensile strength σ_t and uniaxial compressive strength σ_c. The conditions of the experiment are

$$\sigma_1 = \sigma_t, \ \sigma_2 = \sigma_3 = 0 \tag{3.3a}$$
$$\sigma_1 = \sigma_2 = 0, \ \sigma_3 = -\sigma_c \tag{3.3b}$$

By substituting Eqs. 3.2a, 3.2b and 3.3a into Eq. 3.1a and substituting Eqs. 3.2a, 3.2c and 3.3b into Eq. 3.1b, the material constants β and C can be determined as follows:

$$\beta = \frac{\sigma_c - \sigma_t}{\sigma_c + \sigma_t} = \frac{1 - \alpha}{1 + \alpha} \tag{3.4a}$$

$$C = \frac{2\sigma_c\sigma_t}{\sigma_c + \sigma_t} = \frac{2}{1+\alpha}\sigma_t \qquad (3.4b)$$

By substituting Eqs. 3.2a, 3.2b and 3.4 into Eq. 3.1a and substituting Eqs. 3.2a, 3.2c and 3.4 into Eq. 3.1b, the Yu-UST in terms of the principal stresses can be obtained as follows:

$$F = \sigma_1 - \frac{\alpha}{1+b}(b\sigma_2 + \sigma_3) = \sigma_t, \text{ when } \sigma_2 \le \frac{\sigma_1 + \alpha\sigma_3}{1+\alpha} \qquad (3.5a)$$

$$F' = \frac{1}{1+b}(\sigma_1 + b\sigma_2) - \alpha\sigma_3 = \sigma_t, \text{ when } \sigma_2 \ge \frac{\sigma_1 + \alpha\sigma_3}{1+\alpha} \qquad (3.5b)$$

The Yu-UST gives a series of failure criteria and establishes the relationship among various failure criteria.

Obviously, the Yu-UST is physically meaningful and has a unified mechanical model, it has a simple and linear unified mathematical expression, which can be suitable for different kinds of materials.

The Yu-UST can be easily extended and developed. It also has rich and varied contents and agrees with existing data of experiments.

3.3   Special cases of the Yu-UST for different parameter *b*

The Yu-UST is a set of failure criteria. Many failure criteria can be introduced by varying parameter b. The five special cases of the Yu-UST with $b = 0$, $b = 1/4$, $b = 1/2$, $b = 3/4$ and $b = 1$ are discussed as follows:

1   When $b = 0$, the Mohr-Coulomb strength theory can be deduced from the Yu-UST as follows:

$$F = F' = \sigma_1 - \alpha\sigma_3 = \sigma_t \qquad (3.6a)$$

$$F = F' = \frac{1}{\alpha}\sigma_1 - \sigma_3 = \sigma_c \qquad (3.6b)$$

2   When $b = 1/4$, a new failure criterion is deduced from the Yu-UST as follows:

$$F = \sigma_1 - \frac{\alpha}{5}(\sigma_2 + 4\sigma_3) = \sigma_t, \text{ when } \sigma_2 \le \frac{\sigma_1 + \alpha\sigma_3}{1+\alpha} \qquad (3.7a)$$

$$F' = \frac{1}{5}(4\sigma_1 + \sigma_2) - \alpha\sigma_3 = \sigma_t, \text{ when } \sigma_2 \ge \frac{\sigma_1 + \alpha\sigma_3}{1+\alpha} \qquad (3.7b)$$

3   When $b = 1/2$, a new failure criterion is deduced from the Yu-UST as follows:

$$F = \sigma_1 - \frac{\alpha}{3}(\sigma_2 + 2\sigma_3) = \sigma_t, \text{ when } \sigma_2 \leq \frac{\sigma_1 + \alpha\sigma_3}{1+\alpha} \quad (3.8a)$$

$$F' = \frac{1}{3}(2\sigma_1 + \sigma_2) - \alpha\sigma_3 = \sigma_t, \text{ when } \sigma_2 \geq \frac{\sigma_1 + \alpha\sigma_3}{1+\alpha} \quad (3.8b)$$

4   When $b = 3/4$, a new failure criterion is deduced from the Yu-UST as follows:

$$F = \sigma_1 - \frac{\alpha}{7}(3\sigma_2 + 4\sigma_3) = \sigma_t, \text{ when } \sigma_2 \leq \frac{\sigma_1 + \alpha\sigma_3}{1+\alpha} \quad (3.9a)$$

$$F' = \frac{1}{7}(4\sigma_1 + 3\sigma_2) - \alpha\sigma_3 = \sigma_t, \text{ when } \sigma_2 \geq \frac{\sigma_1 + \alpha\sigma_3}{1+\alpha} \quad (3.9b)$$

5   When $b = 1$, a new failure criterion is deduced from the Yu-UST as follows:

$$F = \sigma_1 - \frac{\alpha}{2}(\sigma_2 + \sigma_3) = \sigma_t, \text{ when } \sigma_2 \leq \frac{\sigma_1 + \alpha\sigma_3}{1+\alpha} \quad (3.10a)$$

$$F' = \frac{1}{2}(\sigma_1 + \sigma_2) - \alpha\sigma_3 = \sigma_t, \text{ when } \sigma_2 \geq \frac{\sigma_1 + \alpha\sigma_3}{1+\alpha} \quad (3.10b)$$

This is the twin-shear strength theory proposed by Mao-Hong Yu (Yu et al., 1985).

3.4 Limit loci of the Yu-UST by varying parameter b in the π-plane

The relationships between the coordinates of the deviatoric plane (x, y, z) and the principal stresses $(\sigma_1, \sigma_2, \sigma_3)$ are

$$\sigma_1 = \frac{1}{3}\left(\sqrt{6}y + \sqrt{3}z\right)$$

$$\sigma_2 = \frac{1}{6}\left(2\sqrt{3}z - \sqrt{6}y - 3\sqrt{2}x\right) \quad (3.11)$$

$$\sigma_3 = \frac{1}{6}\left(3\sqrt{2}x - \sqrt{6}y + 2\sqrt{3}z\right)$$

Substituting Eq. 3.11 into Eqs. 3.5a and 3.5b, the equations of the Yu-UST in the deviatoric plane when $-\frac{\sqrt{3}}{3}y \leq x \leq 0$ ($\sigma_1 \geq \sigma_2 \geq \sigma_3$ or $0 \leq \theta \leq 60°$) can be obtained as

$$F = -\frac{\sqrt{2}(1-b)}{2(1+b)}\alpha x + \frac{\sqrt{6}(2+\alpha)}{6}y + \frac{\sqrt{3}(1-\alpha)}{3}z = \sigma_t$$

(3.12a)

$$\text{when } x \geq -\frac{\sqrt{3}}{1+2\alpha}y$$

$$F' = -\left(\frac{b}{1+b}+\alpha\right)\frac{\sqrt{2}}{2}x + \left(\frac{2-b}{1+b}+\alpha\right)\frac{\sqrt{6}}{6}y + \frac{\sqrt{3}(1-\alpha)}{3}z = \sigma_t$$

(3.12b)

$$\text{when } x \leq -\frac{\sqrt{3}}{1+2\alpha}y$$

A great number of new failure criteria can be generated from the Yu-UST by changing α and b. The general shape of the limit loci of the Yu-UST on the deviatoric plane is shown in Figure 3.3.

Material parameters α and σ_t are the tension–compression strength ratio and the uniaxial tensile strength, respectively, and b is a material parameter that reflects the influence of intermediate principal shear stress. A series of limit surfaces can be obtained by varying b.

Five special cases will be discussed with values of $b = 0$, $b = 1/4$, $b = 1/2$, $b = 3/4$ and $b = 1$.

1   Substituting $b = 0$ into Eq. 3.12a and 3.12b, an equation can be obtained as follows:

$$F = F' = -\frac{\sqrt{2}}{2}\alpha x + \frac{\sqrt{6}}{6}(2+\alpha)y + \frac{\sqrt{3}}{3}(1-\alpha)z = \sigma_t$$

(3.13)

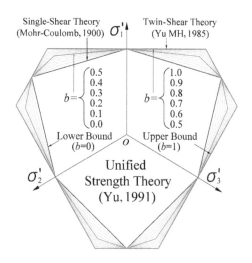

Figure 3.3 A serials of yield loci of the Yu-UST

This is the Mohr-Coulomb strength theory. The limit locus of the Mohr-Coulomb strength theory is the lower bound of the convex limit loci.

2   Substituting $b = 1/4$ into Eq. 3.12a and 3.12b, a new equation can be obtained as follows:

$$F = -\frac{3\sqrt{2}}{10}\alpha x + \frac{\sqrt{6}}{6}(2+\alpha)y + \frac{\sqrt{3}}{3}(1-\alpha)z = \sigma_t$$

$$\text{when } x \geq -\frac{\sqrt{3}}{1+2\alpha}y$$

(3.14a)

$$F' = -\left(\frac{1}{5}+\alpha\right)\frac{\sqrt{2}}{2}x + \left(\frac{7}{5}+\alpha\right)\frac{\sqrt{6}}{6}y + \frac{\sqrt{3}}{3}(1-\alpha)z = \sigma_t$$

$$\text{when } x \leq -\frac{\sqrt{3}}{1+2\alpha}y$$

(3.14b)

This is the limit surface of a new failure criterion.

3   When $b = 1/2$, substituting it into Eqs. 3.12a and 3.12b, we have the new equation as follows:

$$F = -\frac{\sqrt{2}}{6}\alpha x + \frac{\sqrt{6}}{6}(2+\alpha)y + \frac{\sqrt{3}}{3}(1-\alpha)z = \sigma_t$$

$$\text{when } x \geq -\frac{\sqrt{3}}{1+2\alpha}y$$

(3.15a)

$$F' = -\left(\frac{1}{3}+\alpha\right)\frac{\sqrt{2}}{2}x + (1+\alpha)\frac{\sqrt{6}}{6}y + \frac{\sqrt{3}}{3}(1-\alpha)z = \sigma_t$$

$$\text{when } x \leq -\frac{\sqrt{3}}{1+2\alpha}y$$

(3.15b)

This is a new failure criterion. It is intermediate between the Mohr-Coulomb strength theory and the twin-shear strength theory.

4   When $b = 3/4$, substituting it into Eqs. 3.12a and 3.12b, we have the new equation as follows:

$$F = -\frac{\sqrt{2}}{14}\alpha x + \frac{\sqrt{6}}{6}(2+\alpha)y + \frac{\sqrt{3}}{3}(1-\alpha)z = \sigma_t$$

$$\text{when } x \geq -\frac{\sqrt{3}}{1+2\alpha}y$$

(3.16a)

$$F' = -\left(\frac{3}{7}+\alpha\right)\frac{\sqrt{2}}{2}x + \left(\frac{5}{7}+\alpha\right)\frac{\sqrt{6}}{6}y + \frac{\sqrt{3}}{3}(1-\alpha)z = \sigma_t$$

(3.16b)

$$\text{when } x \le -\frac{\sqrt{3}}{1+2\alpha}y$$

This is the limit surface of a new failure criterion.

5 When $b = 1$, substituting it into Eq. 3.12a and 3.12b we have the new equation as follows:

$$F = \frac{\sqrt{6}}{6}(2+\alpha)y + \frac{\sqrt{3}}{3}(1-\alpha)z = \sigma_t$$

(3.17a)

$$\text{when } x \ge -\frac{\sqrt{3}}{1+2\alpha}y$$

$$F' = -\left(\frac{1}{2}+\alpha\right)\frac{\sqrt{2}}{2}x + \left(\frac{1}{2}+\alpha\right)\frac{\sqrt{6}}{6}y + \frac{\sqrt{3}}{3}(1-\alpha)z = \sigma_t$$

(3.17b)

$$\text{when } x \le -\frac{\sqrt{3}}{1+2\alpha}y$$

This is the twin-shear strength theory proposed by Yu *et al.* (1985). The limit locus of the twin-shear strength theory is the upper bound of the convex area as shown in Figure 3.3. A great number of failure criteria can be deduced from the Yu-UST.

3.5   Variation of limit loci of the Yu-UST when $\alpha = 1/2$

The Yu-UST can also be adopted for the materials with different tension–compression strength ratios.

If the tensile strength of materials is half its compressive strength, that is, the tension–compression strength ratio is $\alpha = \sigma_t/\sigma_c = 1/2$, then the corresponding equation of the Yu-UST is given by

$$F = -\frac{(1-b)\sqrt{2}}{4(1+b)}x + \frac{5\sqrt{6}}{12}y + \frac{\sqrt{3}}{6}z = \sigma_t, \text{ when } x \ge -\frac{\sqrt{3}}{2}y \quad (3.18a)$$

$$F' = -\frac{(3b+1)\sqrt{2}}{4(1+b)}x + \frac{(5-b)\sqrt{6}}{12(1+b)}y + \frac{\sqrt{3}}{6}z = \sigma_t,$$

$$\text{when } x \le -\frac{\sqrt{3}}{2}y \quad (3.18b)$$

The equations of the failure criteria of the Yu-UST for materials for which $\alpha = 1/2$ on the deviatoric plane with $b = 0$, $b = 1/4$, $b = 1/2$, $b = 3/4$ and $b = 1$ are described as follows:

1   Materials for which $\alpha = \sigma_t/\sigma_c = 1/2$ and $b = 0$:

The failure criterion for $\alpha = 1/2$ materials on the deviatoric plane with $b = 0$ can be obtained from the Yu-UST. This is identical with the Mohr-Coulomb strength theory. The equation is

$$F = F' = -\frac{\sqrt{2}}{4}x + \frac{5\sqrt{6}}{12}y + \frac{\sqrt{3}}{6}z = \sigma_t \tag{3.19}$$

2   Materials for which $\alpha = \sigma_t/\sigma_c = 1/2$ and $b = 1/4$:

The failure criterion for $\alpha = 1/2$ materials on the deviatoric plane with $b = 1/4$ can also be obtained from the Yu-UST, which gives a new criterion.

$$F = -\frac{3\sqrt{2}}{20}x + \frac{5\sqrt{6}}{12}y + \frac{\sqrt{3}}{6}z = \sigma_t, \text{ when } x \geq -\frac{\sqrt{3}}{2}y \tag{3.20a}$$

$$F' = -\frac{7\sqrt{2}}{20}x + \frac{19\sqrt{6}}{60}y + \frac{\sqrt{3}}{6}z = \sigma_t, \text{ when } x \leq -\frac{\sqrt{3}}{2}y \tag{3.20b}$$

3   Materials for which $\alpha = \sigma_t/\sigma_c = 1/2$ and $b = 1/2$:

The failure criterion for $\alpha = 1/2$ materials on the deviatoric plane with $b = 1/2$ can be obtained from the Yu-UST, which gives a new criterion.

$$F = -\frac{\sqrt{2}}{12}x + \frac{5\sqrt{6}}{12}y + \frac{\sqrt{3}}{6}z = \sigma_t, \text{ when } x \geq -\frac{\sqrt{3}}{2}y \tag{3.21a}$$

$$F' = -\frac{5\sqrt{2}}{12}x + \frac{\sqrt{6}}{4}y + \frac{\sqrt{3}}{6}z = \sigma_t, \text{ when } x \leq -\frac{\sqrt{3}}{2}y \tag{3.21b}$$

4   Materials for which $\alpha = \sigma_t/\sigma_c = 1/2$ and $b = 3/4$:

The failure criterion for $\alpha = 1/2$ materials on the deviatoric plane with $b = 3/4$ can be obtained from the Yu-UST; this yields

$$F = -\frac{\sqrt{2}}{28}x + \frac{5\sqrt{6}}{12}y + \frac{\sqrt{3}}{6}z = \sigma_t, \text{ when } x \geq -\frac{\sqrt{3}}{2}y \tag{3.22a}$$

$$F' = -\frac{13\sqrt{2}}{28}x + \frac{17\sqrt{6}}{84}y + \frac{\sqrt{3}}{6}z = \sigma_t, \text{ when } x \leq -\frac{\sqrt{3}}{2}y \tag{3.22b}$$

5   Materials for which $\alpha = \sigma_t/\sigma_c = 1/2$ and $b = 1$:

The failure criterion for materials for which $\alpha = 1/2$ on the deviatoric plane with $b = 1$ can be obtained from the Yu-UST. This is the twin-shear strength theory (Yu *et al.*, 1985). The failure locus on the deviatoric plane is the upper bound of the convex limit loci.

$$F = \frac{5\sqrt{6}}{12}y + \frac{\sqrt{3}}{6}z = \sigma_t, \text{ when } x \geq -\frac{\sqrt{3}}{2}y \qquad (3.23a)$$

$$F' = -\frac{\sqrt{2}}{2}x + \frac{\sqrt{6}}{6}y + \frac{\sqrt{3}}{6}z = \sigma_t, \text{ when } x \leq -\frac{\sqrt{3}}{2}y \qquad (3.23b)$$

As discussed earlier, the Yu-UST gives a series of new yield and failure criteria, establishes a relationship among various failure criteria and encompasses previous yield criteria and failure criteria as its special cases or linear approximations. In particular, the Yu-UST with $b = 1/2$ and $b = 3/4$ can serve as new criteria, which can conveniently replace the smooth-ridge models.

3.6   Limit surfaces of the Yu-UST in principal stress space

The limit surfaces of strength theories can be illustrated in the principal stress space (Haigh, 1920; Westergaard, 1920). The principal stress space is also called the Haigh-Westergaard space. The limit surfaces of the Yu-UST are usually a semi-infinite hexagonal cone with unequal sides and a dodecahedron cone with unequal sides, as shown in Figure 3.4. The shape and size of the limiting hexagonal cone depend on

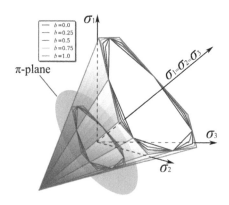

Figure 3.4 Limit surface of the unified strength theory

the parameter b and the tension–compression strength ratio α. If $\alpha = 1$ and $b = 0$, the limit surface will be reduced to an infinite hexagonal cylindrical surface with size equal to the yield surface of the Tresca yield criterion. If $a = b = 1$, the limit surface will be reduced to an infinite hexagonal cylindrical surface with size equal to the yield surface of the twin-shear yield criterion.

In engineering practice, the compressive strength of materials σ_c is often much greater than the tensile strength σ_t, because a region in tension becomes smaller, while it becomes larger in compression.

The limiting surfaces of the Mohr-Coulomb criterion and Yu-UST with different values of b ($b = 0$, $b = 1/4$, $b = 1/2$, $b = 3/4$ and $b = 1$, respectively) are shown in Figure 3.5.

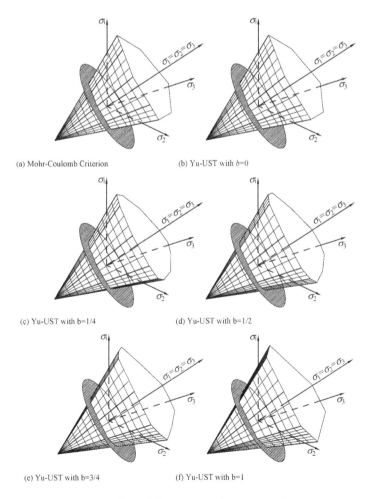

(a) Mohr-Coulomb Criterion

(b) Yu-UST with $b=0$

(c) Yu-UST with $b=1/4$

(d) Yu-UST with $b=1/2$

(e) Yu-UST with $b=3/4$

(f) Yu-UST with $b=1$

Figure 3.5 Limit surface of the Yu-UST (from Yu, 1998)

3.7   Limit loci of the Yu-UST in plane stress state

The limit loci of the Yu-UST in the plane stress state are the intersection line of the limit surface in principal stress space and the $\sigma_1 - \sigma_2$ plane. Its shape and size depend on the values of b and α. It will be transformed into hexagon when $b = 0$ or $b = 1$ and into dodecagon when $0 < b < 1$.

The equations of the 12 limiting loci of the Yu-UST in the plane stress state can be given as follows:

$$\sigma_1 - \frac{\alpha b}{1+b}\sigma_2 = \sigma_t, \quad \frac{\alpha}{1+b}(\sigma_1 + b\sigma_2) = \sigma_t \tag{3.24a}$$

$$\sigma_2 - \frac{\alpha b}{1+b}\sigma_1 = \sigma_t, \quad \frac{\alpha}{1+b}(\sigma_2 + b\sigma_1) = \sigma_t \tag{3.24b}$$

$$\sigma_1 - \frac{\alpha}{1+b}\sigma_2 = \sigma_t, \quad \frac{1}{1+b}\sigma_1 - \alpha\sigma_2 = \sigma_t \tag{3.24c}$$

$$\sigma_2 - \frac{\alpha}{1+b}\sigma_1 = \sigma_t, \quad \frac{1}{1+b}\sigma_2 - \alpha\sigma_1 = \sigma_t \tag{3.24d}$$

$$\frac{\alpha}{1+b}(b\sigma_1 + \sigma_2) = -\sigma_t, \quad \frac{b}{1+b}\sigma_1 - \alpha\sigma_2 = \sigma_t \tag{3.24e}$$

$$\frac{\alpha}{1+b}(b\sigma_2 + \sigma_1) = -\sigma_t, \quad \frac{b}{1+b}\sigma_2 - \alpha\sigma_1 = \sigma_t \tag{3.24f}$$

A series of new failure criteria and new limit loci in the plane stress state can be obtained from the Yu-UST.

The limit loci of the Yu-UST in the plane stress state with different values of b are shown in Figure 3.6 and Figure 3.7 with $\alpha = 0.6$ and $\alpha = 0.3$, respectively.

The limit loci of the Yu-UST in the plane stress state with different values of α are shown in Figures 3.6 and 3.7. Figure 3.6 shows the

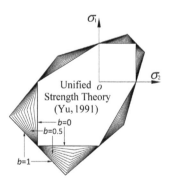

Figure 3.6 Limit loci of the Yu-UST in plane stress state ($\alpha = 0.6$)

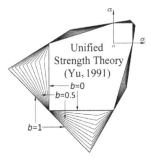

Figure 3.7 Limit loci of the Yu-UST in the plane stress state ($\alpha = 0.3$)

limiting line of the Yu-UST in the $\sigma_1 - \sigma_2$ plane with $\alpha = 0.6$. Figure 3.7 shows the limiting line of the Yu-UST in the $\sigma_1 - \sigma_2$ plane with $\alpha = 0.3$.

If the tensile strength is identical to the compressive strength, the Yu-UST will be transformed into the Yu-unified yield criterion (Yu-UYC).

Various limit loci of the Yu-UST in the plane stress state are shown in Figure 3.8. The Yu-UYC, the Mohr-Coulomb strength theory, the twin-shear strength theory and a series of new failure criteria can be obtained from the Yu-UST.

3.8   Limit loci of the Yu-UST under the $\sigma - \tau$ combined stress state

Under the $\sigma - \tau$ combined stress state, the three principal stresses are

$$\sigma_1 = \frac{\sigma}{2} + \frac{1}{2}\sqrt{\sigma^2 + 4\tau^2}; \sigma_2 = 0; \sigma_3 = \frac{\sigma}{2} - \frac{1}{2}\sqrt{\sigma^2 + 4\tau^2} \qquad (3.25)$$

By substituting Eq. 3.25 into Eq. 3.5a and 3.5b, the expression and the limit loci of the Yu-UST under the $\sigma - \tau$ combined stress state can be obtained.

$$F = \frac{1+b+\alpha}{2+2b}\sqrt{\sigma^2 + 4\tau^2} + \frac{1+b-\alpha}{2+2b}\sigma = \sigma_t$$

$$\text{when}\quad \frac{1+\alpha}{2}\sigma + \frac{1-\alpha}{2}\sqrt{\sigma^2 + 4\tau^2} \geq 0 \qquad (3.26a)$$

$$F' = \frac{1+\alpha+\alpha b}{2+2b}\sqrt{\sigma^2 + 4\tau^2} + \frac{1-\alpha-\alpha b}{2+2b}\sigma = \sigma_t$$

$$\text{when}\quad \frac{1+\alpha}{2}\sigma + \frac{1-\alpha}{2}\sqrt{\sigma^2 + 4\tau^2} \leq 0 \qquad (3.26b)$$

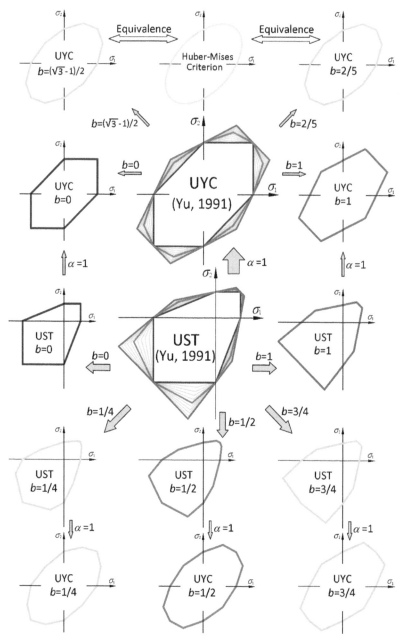

Figure 3.8 Variation of the Yu-UST in the plane stress state

A series of failure criteria of the Yu-UST in the $\sigma - \tau$ stress state can be introduced from the above equations, such as:

1 When $b = 0$, the single-shear strength theory in the $\sigma - \tau$ stress state is given as follows:

$$F = F' = \frac{1+\alpha}{2}\sqrt{\sigma^2 + 4\tau^2} + \frac{1-\alpha}{2}\sigma = \sigma_t \tag{3.27}$$

2 When $b = 1/2$, a new failure criterion in the $\sigma - \tau$ stress state is given as follows:

$$\frac{3+2\alpha}{6}\sqrt{\sigma^2 + 4\tau^2} + \frac{3-2\alpha}{6}\sigma = \sigma_t, \text{ when}$$

$$\frac{1+\alpha}{2}\sigma + \frac{1-\alpha}{2}\sqrt{\sigma^2 + 4\tau^2} \geq 0 \tag{3.28a}$$

$$F' = \frac{2+3\alpha}{6}\sqrt{\sigma^2 + 4\tau^2} + \frac{2-3\alpha}{6}\sigma = \sigma_t, \text{ when}$$

$$\frac{1+\alpha}{2}\sigma + \frac{1-\alpha}{2}\sqrt{\sigma^2 + 4\tau^2} \leq 0 \tag{3.28b}$$

3 When $b = 1$, the twin-shear strength theory in the $\sigma - \tau$ stress state is given as follows:

$$F = \frac{2+\alpha}{4}\sqrt{\sigma^2 + 4\tau^2} + \frac{2-\alpha}{4}\sigma = \sigma_t, \text{ when}$$

$$\frac{1+\alpha}{2}\sigma + \frac{1-\alpha}{2}\sqrt{\sigma^2 + 4\tau^2} \geq 0 \tag{3.29a}$$

$$F' = \frac{1+2\alpha}{4}\sqrt{\sigma^2 + 4\tau^2} + \frac{1-2\alpha}{4}\sigma = \sigma_t, \text{ when}$$

$$\frac{1+\alpha}{2}\sigma + \frac{1-\alpha}{2}\sqrt{\sigma^2 + 4\tau^2} \leq 0 \tag{3.29b}$$

The failure loci of the Yu-UST with $b = 0$, $b = 1/4$, $b = 1/2$, $b = 3/4$ and $b = 1$ in the $\sigma - \tau$ stress state for two kinds of materials with $\alpha = 1$ and $\alpha = 0.4$, respectively, are shown in Figure 3.9.

The limit loci of the Yu-UST with $\alpha = 1$ are identical with the Yu-unified yield criterion (Yu-YUC) under the $\sigma - \tau$ combined stress state. It will be transformed into that of the Tresca yield criterion when $b = 0$, that of the twin-shear stress yield criterion when $b = 1$

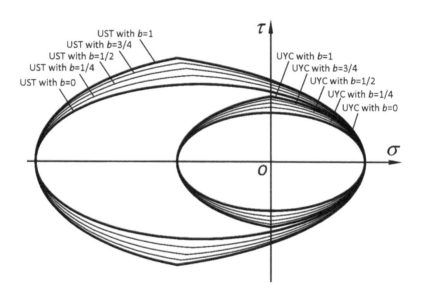

Figure 3.9 Limit loci of the Yu-UST at in the $\sigma - \tau$ combined stress state

and a new yield criterion when $b = 1/2$. The limit loci of the Mohr-Coulomb strength theory, the twin-shear strength theory and several new failure criteria can be obtained when $b = 0$, $b = 1$ and $b = 1/2$, respectively.

3.9   Experiment verification of Yu-UST

The study of the strength of materials under complex stress state is complicated both in theory and in tests. The experimental verification of strength theories is very important, and it can help us to analyze the theory. Many comparisons of the Yu-UST with the experimental data have been given by Yu (2018). These experiments carried out by many researchers are very important. Two experimental verifications (Shibata and Karube, 1965; Takahashi and Koide, 1989) of unified strength theory for soil and rock are shown in Figure 3.10.

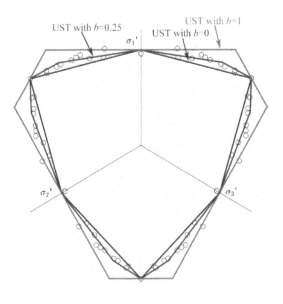

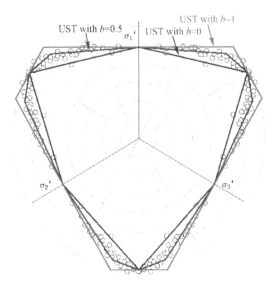

Figure 3.10 Limit loci in the π-plane for (a) normally consolidated soil (b) Shirahama sandstone

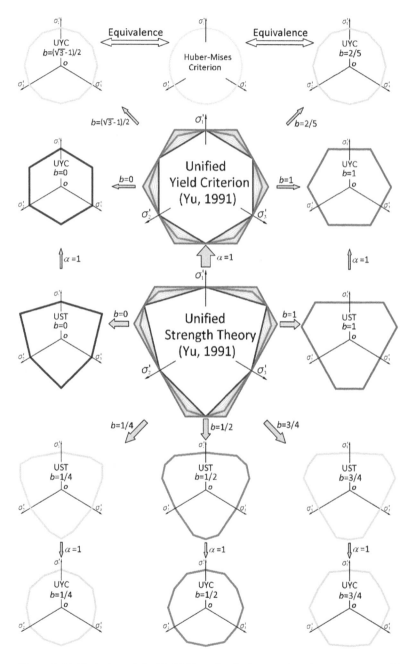

Figure 3.11 Variation of the Yu-UST on the deviatoric plane by varying α and b

3.10 Summary

The unified strength theory is described in this chapter. It is not a single failure criterion but a set of series of failure criteria. It is introduced from the twin-shear idea and twin-shear element. It has two twin-shear elements, two mathematical expressions and an additional condition as follows:

$$F = \sigma_1 - \frac{\alpha}{1+b}(b\sigma_2 + \sigma_3) = \sigma_t, \text{ when } \sigma_2 \leq \frac{\sigma_1 + \alpha\sigma_3}{1+\alpha} \qquad (3.30a)$$

$$F' = \frac{1}{1+b}(\sigma_1 + b\sigma_2) - \alpha\sigma_3 = \sigma_t, \text{ when } \sigma_2 \geq \frac{\sigma_1 + \alpha\sigma_3}{1+\alpha} \qquad (3.30b)$$

The Yu-UST has clear physical meaning and a unified mechanical model. It has simple and linear unified mathematical expression. The series of yield loci of UST cover the whole convex region from the lower bound to upper bound. UST encompasses many well-known yield criteria and failure criteria as its special cases or linear approximations.

A series of limit surface and loci can be obtained from the Yu-UST. The unified strength theory forms an entire spectrum of convex criteria, which can be used for most isotropic materials. The relationships between the Yu-UST and its several special cases are illustrated in Figure 3.11.

References

Haigh, B.T. (1920) The strain energy function and the elastic limit. *Engineering, London*, 109, 158–160.

Shibata, T. & Karube, D. (1965) Influence of the variation of the intermediate principal stress on the mechanical properties of normally consolidated clays. *Proc. Sixth Int. Conf. on Soil Mechanics and Found Engrg*, 1, 359–363.

Takahashi, M. & Koide, H. (1989) Effect of the intermediate principal stress on strength and deformation behavior of sedimentary rocks at the depth shallower than 200m. In: Maury V, Fourmaintraux D (eds) *International Symposium on Rock at Great Depth*, Vol. 1. Balkema, Rotterdam, pp. 19–26.

Westergaard, H.M. (1920) On the resistance of ductile materials to combined stresses in two or three directions perpendicular to one another. *J. Franklin Inst.*, 189, 627–640.

Yu, M.H. (1992) *A New System of Strength Theory*. Xian Jiaotong University Press, Xian, China (in Chinese).

Yu, M.H. (1998) *Twin-Shear Strength Theory and Its Applications*. Science Press, Beijing (in Chinese).

Yu, M.H. (2018) *Unified Strength Theory and Its Applications*, 2nd edition. Springer and Xi'an Jiaotong University Press, Singapore and Xi'an.

Yu, M.H. & He, L.N. (1991) A new model and theory on yield and failure of materials under the complex stress state. In: Jono, M. & Inoue, T. (eds) *Mechanical Behavior of Materials-6 (ICM-6)*, Vol. 3. Pergamon Press, Oxford, pp. 841–846.

Yu, M.H., He, L.N. & Song, L.Y. (1985) Twin shear stress theory and its generalization. *Science in China Series A, English Edition*, 28(11), 1174–1183.

Readings

[**Readings 3–1**] "The development of Unified Strength Theory (UST) is an event in phenomenological material science. The model of UST provides a new family of material models. It contains a number of new models and highlights interrelations between known models. The UST-model can be fitted to different materials and is therefore suitable for the analysis of experimental results. The material parameters can be computed using results of only three experiments (e.g. tension, compression and torsion).

The following advantages of UST are to be pointed out:

1   The concept of understanding the stress components $\sigma_{ij} = (\sigma_i + \sigma_j)/2$ and $\tau_{ij} = (\sigma_i - \sigma_j)/2$ based upon polyhedral elements,
2   Extension of the stress state parameter according to Lode by "twin-shear stress" parameters $\mu_\tau = \tau_{12}/\tau_{13}$, $\mu'_\tau = \tau_{23}/\tau_{13}$, $\mu_\tau + \mu'_\tau = 1$,
3   Fitting of the parameters (b, α) to various measured data found in literature, as well as recommendations for different types of materials,
4   Physical interpretation of the parameters,
5   Incorporation of the third deviatoric invariant I'_3 into the model,
6   Simple computation of the equivalent stress σ_{eq} as well as of the derivative $\partial\sigma_{eq}/\partial\sigma_{ij}$ everywhere except for singular points."

Kolupaev, V.A. & Altenbach, H. (2009) Strength hypotheses of Mao-Hong Yu and its generalisation." In: Kuznetsov S.A. (Hrsg) *2nd Conference Problems in Nonlinear Mechanics of Deformable Solids, 8–11, December 2009*. Kazan State University, Kazan, pp. 10–12.

4 Yu-unified yield criteria for metals

4.1 Introduction

As stated in Chapter 1, the tensile strength and compressive strength of most materials are different. The function of strength theory for these materials is expressed by two material parameters as follows:

$$F(\sigma_1, \sigma_2, \sigma_3; \sigma_t, \sigma_c) = 0 \tag{4.1}$$

This is a two-parameter strength theory. The Yu-UST for these materials has been described in detail in Chapter 3.

On the other hand, some materials have an identical strength both in tension and compression, as shown in Figure 1.1 in Chapter 1, i.e.

$$\sigma_t = \sigma_c = \sigma_y \tag{4.2}$$

In this case, the function of strength theory can be simplified to a one-parameter function as follows:

$$F(\sigma_1, \sigma_2, \sigma_3; \sigma_y) = 0 \tag{4.3}$$

4.2 The Yu-unified yield criterion (Yu-UYC)

The Yu-UST has been described in Chapter 3. The mathematical expressions can be rewritten as follows:

$$F = \sigma_1 - \frac{\alpha}{1+b}(b\sigma_2 + \sigma_3) = \sigma_t, \text{ when } \sigma_2 \leq \frac{\sigma_1 + \alpha\sigma_3}{1+\alpha} \tag{4.4a}$$

$$F' = \frac{1}{1+b}(\sigma_1 + b\sigma_2) - \alpha\sigma_3 = \sigma_t, \text{ when } \sigma_2 \geq \frac{\sigma_1 + \alpha\sigma_3}{1+\alpha} \tag{4.4b}$$

where $\alpha = \sigma_t/\sigma_c$ is the ratio of tensile strength of material and compressive strength of material.

For some kinds of materials, the tensile strength σ_t equals compressive strength σ_c, $\alpha = \sigma_t / \sigma_c = 1$. The Yu-UST is simplified to the Yu-unified yield criterion (Yu and He, 1991; Yu, 2004, 2018). The mathematical expressions are shown as follows:

$$f = \sigma_1 - \frac{1}{1+b}(b\sigma_2 + \sigma_3) = \sigma_y, \text{ when } \sigma_2 \leq \frac{1}{2}(\sigma_1 + \sigma_3) \qquad (4.5a)$$

$$f' = \frac{1}{1+b}(\sigma_1 + b\sigma_2) - \sigma_3 = \sigma_y, \text{ when } \sigma_2 \geq \frac{1}{2}(\sigma_1 + \sigma_3) \qquad (4.5b)$$

These equations can be used for metals and materials having the same tensile strength and compressive strength. It is referred to as the Yu-UYC, where b is a yield criterion parameter that represents the effect of the intermediate principal shear stress on the yield of materials.

4.3 Special cases of the Yu-UYC

A series of yield criteria can be deduced from the Yu-UYC.

1   When $b = 0$, the equation of the Yu-UYC is

$$f = f' = (\sigma_1 - \sigma_3) = \sigma_y \qquad (4.6)$$

This is the well-known maximum shear stress strength theory (i.e. Tresca yield criterion).

2   When $b = 1/4$, the equation of the Yu-UYC is

$$f = \sigma_1 - \frac{1}{5}(\sigma_2 + 4\sigma_3) = \sigma_y, \text{ when } \sigma_2 \leq \frac{1}{2}(\sigma_1 + \sigma_3) \qquad (4.7a)$$

$$f' = \frac{1}{5}(4\sigma_1 + \sigma_2) - \sigma_3 = \sigma_y, \text{ when } \sigma_2 \geq \frac{1}{2}(\sigma_1 + \sigma_3) \qquad (4.7b)$$

This is a new yield criterion.

3   When $b = (\sqrt{3} - 1)/2 = 0.366$, the equation of the Yu-UYC is

$$f = \sigma_1 - (2 - \sqrt{3})[\sigma_2 + (1 + \sqrt{3})\sigma_3] = \sigma_y, \text{when } \sigma_2 \leq \frac{1}{2}(\sigma_1 + \sigma_3) \ (4.8a)$$

$$f' = (2 - \sqrt{3})[1 + \sqrt{3})\sigma_1 + \sigma_2] - \sigma_3 = \sigma_y, \text{when } \sigma_2 \geq \frac{1}{2}(\sigma_1 + \sigma_3) \ (4.8b)$$

4   When $b = 2/5$, the equation of the Yu-UYC is

$$f = \sigma_1 - \frac{1}{7}(5\sigma_2 + 2\sigma_3) = \sigma_y, \text{ when } \sigma_2 \leq \frac{1}{2}(\sigma_1 + \sigma_3) \tag{4.9a}$$

$$f' = \frac{1}{7}(5\sigma_1 + 2\sigma_2) - \sigma_3 = \sigma_y, \text{ when } \sigma_2 \geq \frac{1}{2}(\sigma_1 + \sigma_3) \tag{4.9b}$$

These two criteria are the two new yield criteria. The well-known Huber-Mises yield criterion can be linear approximated by the Yu-UYC with $b = (\sqrt{3} - 1)/2$ or $b = 2/5$.

5   When $b = 1/2$, the equation of the Yu-UYC is

$$f = \sigma_1 - \frac{1}{3}(\sigma_2 + 2\sigma_3) = \sigma_y, \text{ when } \sigma_2 \leq \frac{1}{2}(\sigma_1 + \sigma_3) \tag{4.10a}$$

$$f' = \frac{1}{3}(2\sigma_1 + \sigma_2) - \sigma_3 = \sigma_y, \text{ when } \sigma_2 \geq \frac{1}{2}(\sigma_1 + \sigma_3) \tag{4.10b}$$

This is a new yield criterion.

6   When $b = 3/4$, the equation of the Yu-UYC is

$$f = \sigma_1 - \frac{1}{7}(3\sigma_2 + 4\sigma_3) = \sigma_y, \text{ when } \sigma_2 \leq \frac{1}{2}(\sigma_1 + \sigma_3) \tag{4.11a}$$

$$f' = \frac{1}{7}(4\sigma_1 + 3\sigma_2) - \sigma_3 = \sigma_y, \text{ when } \sigma_2 \geq \frac{1}{2}(\sigma_1 + \sigma_3) \tag{4.11b}$$

This is a new yield criterion.

7   When $b = 1$, the equation of the Yu-UYC is

$$f = \sigma_1 - \frac{1}{2}(\sigma_2 + \sigma_3) = \sigma_y, \text{ when } \sigma_2 \leq \frac{1}{2}(\sigma_1 + \sigma_3) \tag{4.12a}$$

$$f' = \frac{1}{2}(\sigma_1 + \sigma_2) - \sigma_3 = \sigma_y, \text{ when } \sigma_2 \geq \frac{1}{2}(\sigma_1 + \sigma_3) \tag{4.12b}$$

This is the twin-shear yield criterion (Yu, 1961, 1983). This criterion assumes that yielding begins when the sum of the two larger principal shear stresses reaches a critical value C. The mathematical model of the twin-shear yield criterion is

$$f = \tau_{13} + \tau_{12} = C, \text{ when } \tau_{12} \geq \tau_{23} \tag{4.13a}$$

$$f' = \tau_{13} + \tau_{23} = C, \text{ when } \tau_{12} \leq \tau_{23} \tag{4.13b}$$

where τ_{13}, τ_{12} and τ_{23} are the maximum principal shear stress, intermediate principal shear stress and minimum principal shear stress, respectively. The twin-shear yield criterion can be adopted for metallic materials with shear yield stress $\tau_y = 2/3\sigma_y$.

The relationship of the Tresca criterion, the Huber-Mises criterion, the twin-shear criterion and other new criteria with the Yu-UYC are illustrated in Figure 4.1. The Yu-UYC with $b = (\sqrt{3}-1)/2$ or $b = 2/5$ is the linear approximation to the Huber-Mises yield criterion.

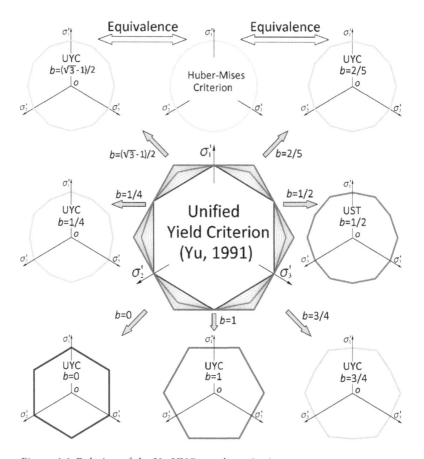

Figure 4.1 Relation of the Yu-UYC to other criteria

4.4 Experimental determination of parameter *b*

The parameter *b* in the Yu-UYC can be determined from the shear yield strength and the tensile yield strength, as shown in Eq. 4.14.

$$b = \frac{2\tau_y - \sigma_y}{\sigma_y - \tau_y} \qquad (4.14)$$

This equation is obtained by using the pure shear stress condition $(\sigma_1 = \tau_y, \sigma_2 = 0, \sigma_3 = -\tau_y)$. Inversely, the shear yield limit τ_y and the ratio of shear yield limit to tensile yield limit can be given as

$$\tau_y = \frac{b+1}{b+2}\sigma_y, \quad \alpha_\tau = \frac{\tau_y}{\sigma_y} = \frac{b+1}{b+2} \qquad (4.15)$$

Many experiments have been done by the researchers. Several results about the tensile yield limit and shear yield limit are summarized in Table 4.1. Some data before 1975 are taken from the historical survey article of Michno and Findley (1976). The parameter *b* of Yu-UYC can be determined by Eq. 4.14.

Table 4.1 Summary and comparison of the yield criteria with experimental results

Researchers	Materials	Specimen		Value of b
Guest (1900)	Steel, brass, etc.	Tubes	0.47	−0.11
Smith (1909)	Mild steel	Solid rods	0.55–0.56	0.22–0.27
Turner (1909, 1911)	Steels	Review work	0.55–0.65	0.22–0.86
Mason (1909)	Mild steel	Tubes	0.64	0.78
Bishop and Hill (1951)	Polycrystals	Tubes	0.54	0.17
Ivey (1961)	Aluminum alloy	Tubes	0.66	0.94
Rogan and Shelton (1969)	Steel	tubes	0.50	0.00
Phillips *et al.* (1972)	Aluminum	Elevated temperature	0.53	0.13
Winstone (1984)	Nickel alloy	Elevated temperature	0.70	1.33
Wu and Yeh (1991)	Aluminum	Tubes	0.58	0.38
Wu and Yeh (1991)	Stainless steel	Tubes	0.66–0.70	0.94–1.33
Ellyin (1993)	Titanium	Tubes	0.62–0.70	0.63–1.33
Ishikawa (1997)	Stainless steel	Tubes	0.60–0.63	0.50–0.70

4.5   Yu-UYC in the plane stress state

The three-dimensional stress state can be deduced to plane stress state if one of the three principal stresses equals zero. The Yu-UYC can be divided into three cases as follows:

1   When $\sigma_1 \geq \sigma_2 > 0$, $\sigma_3 = 0$, the Yu-UYC in the plane stress state is shown as follows:

$$f = \sigma_1 - \frac{b}{1+b}\sigma_2 = \sigma_y, \text{ when } \sigma_2 \leq \frac{1}{2}\sigma_1 \tag{4.16a}$$

$$f' = \frac{1}{1+b}\sigma_1 + \frac{b}{1+b}\sigma_2 = \sigma_y, \text{ when } \sigma_2 \geq \frac{1}{2}\sigma_1 \tag{4.16b}$$

2   When $\sigma_1 \geq 0$, $\sigma_2 = 0$, $\sigma_3 < 0$, the Yu-UYC in plane stress state is shown as follows:

$$f = \sigma_1 - \frac{1}{1+b}\sigma_3 = \sigma_y, \text{ when } \frac{1}{2}(\sigma_1 + \sigma_3) \geq 0 \tag{4.17a}$$

$$f' = \frac{1}{1+b}\sigma_1 - \sigma_3 = \sigma_y, \text{ when } \frac{1}{2}(\sigma_1 + \sigma_3) \geq 0 \tag{4.17b}$$

3   When $\sigma_1 = 0$, $\sigma_2 \geq \sigma_3 < 0$, the Yu-UYC in plane stress state is shown as follows:

$$f = -\frac{1}{1+b}(b\sigma_2 + \sigma_3) = \sigma_y, \text{ when } \sigma_2 \leq \frac{1}{2}\sigma_3 \tag{4.18a}$$

$$f' = \frac{b}{1+b}\sigma_2 - \sigma_3 = \sigma_y, \text{ when } \sigma_2 \geq \frac{1}{2}\sigma_3 \tag{4.18b}$$

In the general case, the Yu-UYC in the plane stress state (σ_1, σ_2) can be expressed by 12 equations as follows:

$$f = \sigma_1 - \frac{b}{1+b}\sigma_2 = \pm\sigma_y, f = \frac{b}{1+b}\sigma_1 - \sigma_2 = \pm\sigma_y \tag{4.19a, b}$$

$$f = \frac{1}{1+b}\sigma_1 + \frac{b}{1+b}\sigma_2 = \pm\sigma_y, f = \frac{b}{1+b}\sigma_1 + \frac{1}{1+b}\sigma_2 = \pm\sigma_y \tag{4.19c, d}$$

$$f = \sigma_1 - \frac{1}{1+b}\sigma_2 = \pm\sigma_y, f = \frac{1}{1+b}\sigma_1 - \sigma_2 = \pm\sigma_y \tag{4.19e, f}$$

These yield equations and yield loci of the Yu-UYC for any value of parameter b can be obtained. For example, the 12 yield equations of the Yu-UYC under the plane stress state when $b = 1/2$ can be given as follows.

$$f_{1,7} = \sigma_1 - \sigma_2 / 3 = \pm\sigma_y, \quad f_{2,8} = 2\sigma_1 + \sigma_2 = \pm\sigma_y \qquad (4.20a, b)$$

$$f_{3,9} = \frac{1}{3}(\sigma_1 + 2\sigma_2) = \pm\sigma_y \quad f_{4,10} = \frac{1}{3}\sigma_1 - \sigma_2 = \pm\sigma_y \qquad (4.20c, d)$$

$$f_{5,11} = \frac{2}{3}\sigma_1 - \sigma_2 = \pm\sigma_y \quad f_{6,12} = \sigma_1 - \frac{2}{3}\sigma_2 = \mp\sigma_y \qquad (4.20e, f)$$

A series of the yield loci of the Yu-UYC in the plane stress state can be given as shown in Figure 4.2. These yield loci cover all the regions of the convex yield criteria.

Varieties of the yield loci of the Yu-UYC in the plane stress states when $b = 0$, $b = 1/4$, $b = 0.366$, $b = 2/5$, $b = 1/2$, $b = 3/4$, and $b = 1$ are shown in Figure 4.3. It can be seen that the yield loci of the Yu-UYC with $b = (\sqrt{3} - 1)/2 = 0.366$ or $b = 2/5$ are the linear approximation of the Huber-Mises yield condition. The two piecewise linear approximations, i.e. Yu-UYC with $b = (\sqrt{3} - 1)/2 = 0.366$ and

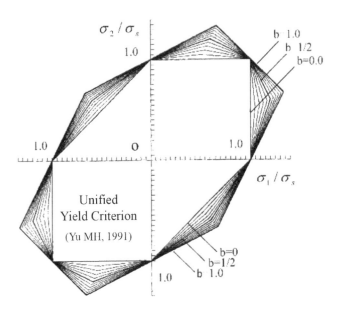

Figure 4.2 Yield loci of the Yu-UYC cover the entire convex region in plane stress (from Yu, 1998)

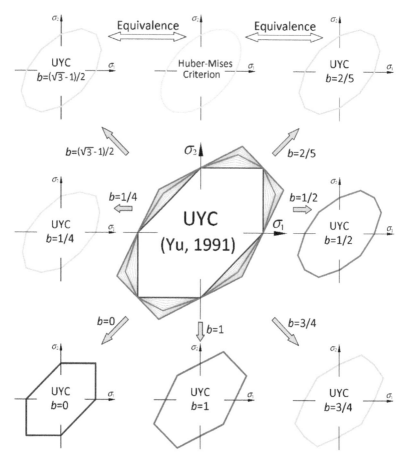

Figure 4.3 Varieties of the Yu-UYC in plane stress

$b = 2/5$, and the Huber-Mises yield criterion are equivalent. The three results obtained by using these three equivalent criterion for structural analyses are almost the same.

4.6 Yu-UYC in the $\sigma - \tau$ stress state

The σ–τ combined stress state is often found in engineering. The three principal stresses can be determined by σ–τ stresses as follows

$$\sigma_1 = \frac{1}{2}(\sigma + \sqrt{\sigma^2 + 4\tau^2}), \ \sigma_2 = 0, \ \sigma_3 = \frac{1}{2}(\sigma - \sqrt{\sigma^2 + 4\tau^2}) \quad (4.21)$$

Substituting Eq. 4.21 into Eq. 4.5 leads to the following expression for the Yu-UYC in $\sigma - \tau$ stress states.

$$f = \frac{2+b}{2+2b}\sqrt{\sigma^2 + 4\tau^2} + \frac{b}{2+2b}\sigma = \sigma_y, \text{ when } \sigma \geq 0 \qquad (4.22a)$$

$$f' = \frac{2+b}{2+2b}\sqrt{\sigma^2 + 4\tau^2} - \frac{b}{2+2b}\sigma = \sigma_y, \text{ when } \sigma \leq 0 \qquad (4.22b)$$

A series of yield criteria in $\sigma - \tau$ stress state can be introduced from the above equation, including

1   When $b = 0$, the single-shear yield criterion in the $\sigma - \tau$ stress state is

$$f = f' = \sqrt{\sigma^2 + 4\tau^2} = \sigma_y \qquad (4.23)$$

2   When $b = 1/2$, the linear Huber-Mises yield criterion in the $\sigma-\tau$ stress is

$$f = \frac{5}{6}\sqrt{\sigma^2 + 4\tau^2} + \frac{1}{6}\sigma = \sigma_y, \text{ when } \sigma \geq 0 \qquad (4.24a)$$

$$f' = \frac{5}{6}\sqrt{\sigma^2 + 4\tau^2} - \frac{1}{6}\sigma = \sigma_y, \text{ when } \sigma \leq 0 \qquad (4.24b)$$

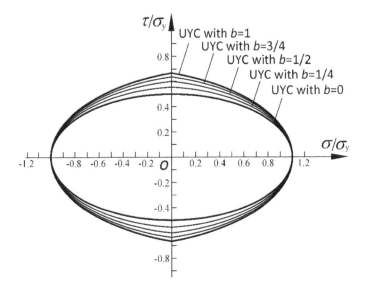

Figure 4.4 Yield loci of the Yu-UYC in the $\sigma - t$ stress state

3   When $b = 1$, the twin-shear yield criterion in the $\sigma - \tau$ stress state is

$$f = \frac{3}{4}\sqrt{\sigma^2 + 4\tau^2} + \frac{1}{4}\sigma = \sigma_y, \text{ when } \sigma \geq 0 \tag{4.25a}$$

$$f' = \frac{3}{4}\sqrt{\sigma^2 + 4\tau^2} - \frac{1}{4}\sigma = \sigma_y, \text{ when } \sigma \leq 0 \tag{4.25b}$$

The yield loci of the Yu-UYC in the $s-\tau$ stress state are shown in Figure 4.5.

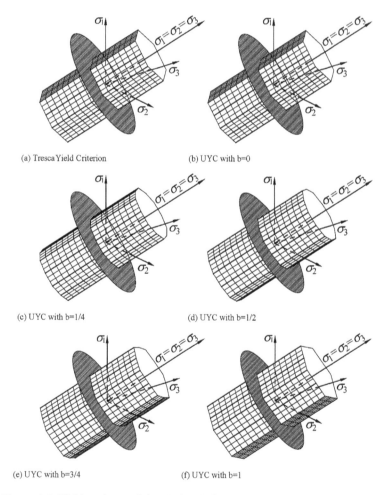

(a) Tresca Yield Criterion

(b) UYC with b=0

(c) UYC with b=1/4

(d) UYC with b=1/2

(e) UYC with b=3/4

(f) UYC with b=1

Figure 4.5 Yield surfaces of the Yu-UYC (from Yu, 1998)

Several experiment results can be seen in Table 4.1.

4.7   Yield surfaces of the Yu-UYC

The yield surface of the Yu-UYC is an infinite hexagonal cylinder (when $b = 0$ or $b = 1$) or an infinite dodecahedron cylinder (when $0 < b < 1$). These yield surfaces are illustrated in Figure 4.6.

Figure 4.6(a) is the yield surface of the Tresca yield criterion. It is the same as the yield surface of the Yu-UYC with $b = 0$.

4.8   Summary

These materials having the identical tensile strength and compressive strength, such as mild steel, are widely used in engineering. The Tresca yield criterion and the Huber-Mises yield criterion are the well-known yield criteria. However, they are single-strength theories and are only suitable for certain kinds of materials. The Tresca yield criterion can only be used for materials with $\sigma_{yt} = \sigma_{yc}$ and $\tau_y = 0.5\sigma_y$. The Huber-Mises yield criterion can only be used for materials with $\sigma_t = \sigma_c$ and $\tau_y = 0.577\sigma_y$. We need a new yield criterion for different mild steel.

A new Yu-UST has been described in Chapter 3. The mathematical expressions are

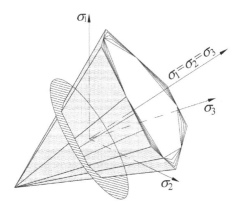

Figure 4.6 A series of limit surfaces of the Yu-UST with various b

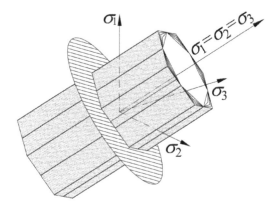

Figure 4.7 A series of limit surfaces of the Yu-UYC with various *b*

$$F = \sigma_1 - \frac{\alpha}{1+b}(b\sigma_2 + \sigma_3) = \sigma_t, \text{ when } \sigma_2 \leq \frac{\sigma_1 + \alpha\sigma_3}{1+\alpha} \qquad (4.26a)$$

$$F' = \frac{1}{1+b}(\sigma_1 + b\sigma_2) - \alpha\sigma_3 = \sigma_t, \text{ when } \sigma_2 \geq \frac{\sigma_1 + \alpha\sigma_3}{1+\alpha} \qquad (4.26b)$$

the ratio $\alpha = \sigma_t / \sigma_c = 1$. The Yu-UST simplified to the Yu-UYC(Yu-UYC) is as follows.

$$f = \sigma_1 - \frac{1}{b}(b\sigma_2 + \sigma_3) = \sigma_y, \text{ when } \sigma_2 \leq \frac{1}{2}(\sigma_1 + \sigma_3) \qquad (4.27a)$$

$$f' = \frac{1}{b}(\sigma_1 + b\sigma_2) - \sigma_3 = \sigma_y, \text{ when } \sigma_2 \geq \frac{1}{2}(\sigma_1 + \sigma_3) \qquad (4.27b)$$

These two set of mathematical expressions are very similar. The only difference is the material parameter α. The yield surface, however, will have many diversifications as shown in Figures 4.6 and 4.7.

References

Bishop, J.F.W. & Hill, R. (1951) A theory of the plastic distortion of a polycrystalline aggregate under combined stresses. *Phil. Mag.*, 42, 414–427.

Ellyin, F. (1993) On the concept of initial and subsequent yield loci. In: Boehler, J.P. (ed) *Failure Criteria of Structured Media*. Balkema, Rotterdam, pp. 293–304.

Guest, J.J. (1900) On the strength of ductile materials under combined stress. *Phil. Mag. and J. Sci.*, 69–133.

Ishikawa, H. (1997) Subsequent yield surface probed from its current center. *Int. J. of Plasticity*, 13(6–7), 533–549.

Ivey, H.J. (1961) Plastic stress-strain relations and yield surfaces for aluminium alloys. *J. Mech. Eng. Sci.*, 3, 15–31.

Mason, W. (1909) Mild steel tubes in compression and under combined stresses. *Proc. Instn. Mech. Engrs.*, 4, 1205.

Michno, M.J. & Findley, W.N. (1973) Experiments to determine small offset yield surfaces of 304L stainless steel under combined tension and torsion. *Acta Mechanic*, 18(3–4), 163–179.

Phillips, A., Liu, C.S. & Justusson, J.W. (1972) An experimental investigation of yield surfaces at elevated temperatures. *Acta Mechanical*, 4, 119–146.

Rogan, H. & Shelton, A. (1969) Effect of pre-stress on the yield and flow of En 25 steel. *J. Strain Analysis*, 4, 138–161.

Smith, C.A. (1909a) Compound stress experiments. *Proc. Instn. Mech. Engrs*, 4, 1237.

Smith, C.A. (1909b) Some experiments on solid steel bars under combined stress. *Engineering*, 20, 238–243.

Turner, L.B. (1909, 1911) The elastic breakdown of materials submitted to compound stresses. *Engineering*, 87, 169 (1909); 92, 115 (1911).

Winstone, M.R. (1984) Influence of prestress on the yield surface of the cast nickel superalloy Mar-M002 at elevated temperature. In: Carlsson, J. & Ohlson, N.G. (eds) *Mechanical Behaviour of Materials-4 (ICM-4)*, Vol. 1. Oxford, Pergamon Press, pp. 199–205.

Wu, H.C. & Yeh, W.C. (1991) On the experimental determination of yield surface and some results of annealed 304 stainless steel. *Int. J. of Plasticity*, 7, 803.

Yu, M.H. (1961) General behavior of isotropic yield function. Res. Report of Xi'an Jiaotong University, Xi'an, China (in Chinese).

Yu, M.H. (1983) Twin shear stress yield criterion. *Int. J. of Mech. Sci.*, 25, 71–74.

Yu, M.H. (1998) *Twin-Shear Strength Theory and Its Applications*. Science Press, Beijing (in Chinese).

Yu, M.H. (2004) *Unified Strength Theory and Its Applications*. Springer, Berlin.

Yu, M.H. (2018) *Unified Strength Theory and Its Applications*, 2nd edition. Springer and Xi'an Jiaotong University Press, Singapore and Xi'an.

Yu, M.H. & He, L.N. (1991) A new model and theory on yield and failure of materials under the complex stress state. In: Jono, M. & Inoue, T. (eds) *Mechanical Behavior of Materials-6 (ICM-6)*, Vol. 3. Pergamon Press, Oxford, pp. 841–846.

Readings

[Readings 4–1] "It should be noted that the parameter b plays an important role in the UYC. It reflects the influence of the intermediate principal stress on the failure of a material. On the other hand, it builds a bridge among different yield criteria. It is this parameter that distinguishes one criterion from another. Hence, the UYC is not a single yield criterion but a theoretical system including a series of regular yield criteria, and it can be applied to various kinds of non-SD materials. In practice, when basic material parameters are obtained by experiments, the value of b can be determined. Whenever parameter b is obtained, the yield criterion for this sort of material is determined and the application is possible. Consequently, b can be regarded as a parameter by which the suitable yield criterion for material of interest can be determined.

"The UYC is a series of piecewise linear yield criteria on the π-plane. The exact form of expression depends on the choice of parameter b. With different choices of parameter b, the UYC can be simplified to the Tresca yield criterion ($b = 0$), the linear approximations of Mises yield criterion ($b = 1/(1 + \sqrt{3})$), the twin-shear yield criterion (TS) ($b = 1$), and a series of new yield criteria. The value of b ranges from 0 to 1 to ensure the yield surfaces are convex. In the stress space, the lower and upper bounds of the yield surfaces on the p-plane, which comply with Druck's convex postulate, are special cases of the UYC, i.e. $b = 0$ for the Tresca, and $b = 1$ for the TS. When the parameter b varies in the range of $0 \leq b \leq 1$, a series of convex yield surfaces between the two limiting surfaces can be obtained, which are suitable for different kinds of non-SD materials."

Wang, L.Z. & Zhang, Y.Q. (2011) Plastic collapse analysis of thin-walled pipes based on unified yield criterion. *International Journal of Mechanical Science*, 53, 348–354.

5 Applications of the Yu-UST and Yu-UYC

5.1 Introduction

The strength theory has been widely used in engineering, including mechanical engineering and civil engineering. It plays an important role in the strength design of structures. More than 200 yield criteria have been proposed. Among them, the Tresca criterion, Huber-Mises criterion and Mohr-Coulomb criterion might be the most commonly used. However, each of them can only be used for a certain kinds of material. Unified strength theory, which can be adopted for various kinds of materials, can describe the strength character more precisely. It has been widely used in many areas, such as natural gas pipeline, water pipe, petroleum pipeline, heat exchange tube in boiler and main steam pipe in power plant. The main drive shaft, the thin-walled pressure tube and the thick-walled cylinder under the internal pressure are the most commonly used simple structures in solid mechanics and engineering. The Yu-UST can be simply applied to their strength analysis and strength design.

In this chapter, we will describe the applications of the unified strength theory in these structures. A series of results can be derived in these problems by using the unified strength theory. These results can be adopted for many kinds of materials. The examples we choose in this chapter can be easily understood by the undergraduate and graduate students who have basic knowledge of the mechanics of materials and students who are studying the mechanics of materials.

5.2 Strength design of a thin-walled pressure cylinder

Suppose that a thin-walled cylinder pressure vessel is subjected to a uniform internal pressure p, as shown in Figure 5.1. The diameter of the cylinder is D, and the wall thickness is t. The material has an axial tensile yield strength of σ_t and an axial compressive yield strength of σ_c. The factor of safety against yielding is to be n. For the purpose of

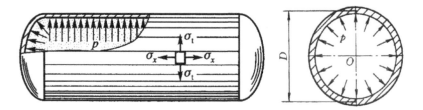

Figure 5.1 Stress state of a thin-walled pressure tube cylinder

engineering design, we need to estimate the elastic limit pressure for a thin-walled vessel.

The stresses of a thin-walled pressure tube are:

$$\sigma_1 = \frac{pD}{2t}, \ \sigma_2 = \frac{pD}{4t}, \ \sigma_3 = 0 \tag{5.1}$$

The expression for the Yu-UST is:

$$F = \sigma_1 - \frac{\alpha}{1+b}(b\sigma_2 + \sigma_3) = \sigma_t, \text{ when } \sigma_2 \le \frac{\sigma_1 + \alpha\sigma_3}{1+\alpha} \tag{5.2a}$$

$$F' = \frac{1}{1+b}(\sigma_1 + b\sigma_2) - \alpha\sigma_3 = \sigma_t, \text{ when } \sigma_2 \ge \frac{\sigma_1 + \alpha\sigma_3}{1+\alpha} \tag{5.2b}$$

It is noted that the stress condition equation of a thin-walled cylinder, Eq. 5.1, satisfies $\sigma_2 = \frac{1}{2}(\sigma_1 + \sigma_3) \le \frac{\sigma_1 + \alpha\sigma_3}{1+\alpha}$. Therefore, we use the first equation of the Yu-UST, Eq. 5–2a. Substituting Eq. 5.1 into Eq. 5.2a, we can get the equation

$$F = \sigma_1 - \frac{\alpha}{1+b}(b\sigma_2 + \sigma_3) = \frac{pD}{2t} - \frac{\alpha b}{1+b}\frac{pD}{4t} = \sigma_t \tag{5.3}$$

The elastic limit pressure of the thin-walled cylinder can be obtained as follows:

$$p_e = \frac{1+b}{2+2b-\alpha b}\frac{4t}{D}\sigma_t, \ [p] = \frac{1+b}{2+2b-\alpha b}\frac{4t}{D}[\sigma], \ [\sigma] = \frac{\sigma_t}{n} \tag{5.4}$$

It can be seen that the application of the Yu-UST in mechanics of materials is very simple. For the SD materials with different tensile–compressive strength ratio α, the relationships between the limit pressure p and the material parameter b of Yu-UST are shown in Figure 5.2.

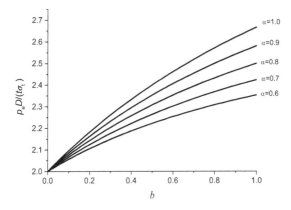

Figure 5.2 The relationships between the limit pressure and the parameter b of the Yu-UST

Table 5.1 The allowable pressures of different strength theories

Strength theory	Allowable pressure [p] (MPa)	Pressure increases compared with the Tresca yield criterion (%)	Pressure increases compared with the Huber-Mises yield criterion (%)
Yu-UYC, $b = 0$	1.33	0	
Tresca yield criterion	1.33	0	
Yu-UYC, $b = 1/4$	1.48	11.3%	
Yu-UYC, $b = (\sqrt{3}-1)/2$	1.54	15.8%	
Huber-Mises yield criterion	1.54	15.8%	0
Yu-UYC, $b = 0.4$	1.56	17.3%	1.6%
Yu-UYC, $b = 1/2$	1.60	20.3%	3.9%
Yu-UYC, $b = 3/4$	1.70	30.0%	10.4%
Yu-UYC, $b = 1.0$	1.78	33.8%	15.6%

If the material of the cylinder is mild steel, which is considered a typical non-SD material, the equations above can be simplified as

$$p_e = \frac{1+b}{2+b}\frac{4t}{D}\sigma_t, \quad [p] = \frac{1+b}{2+b}\frac{4t}{D}[\sigma] \qquad (5.5)$$

If the diameter of cylinder is $D = 200$cm, the yield limit of the mild steel is $\sigma_y = 200$MPa, the wall thickness is $t = 10$mm and the safety factor is $n = 1.5$, the elastic limit pressure of different strength theory can be obtained by Eq. 5.5, as shown in Table 5.1.

Several conclusions can be deduced from the results:

1   It seems that the choice of strength theory has a significant influence on the results of a thin-walled cylinder under internal pressure. When $b = 0$ (Tresca criterion), the elastic limit pressure is the minimum; when $b = 1$ (Twin shear criterion), the elastic limit load is the maximum. It is noted that the different ratios of elastic limit pressure obtained by Yu-UYC with $b = 0$ to that with $b = 1$ is 33.3%.

2   The result obtained by Yu-UYC with $b = 0$ is exactly the same as the result obtained by Tresca yield criterion. Therefore, the result obtained by Tresca yield criterion can be deduced by using Yu-UYC with $b = 0$.

3   The result of the Yu-UYC when $b = (\sqrt{3} - 1)/2 = 0.366$ is the same as the result of the Huber-Mises yield criterion in this case. The Huber-Mises yield criterion can be approximated by Yu-UYC with $b = 0.366$.

4   For convenience, we can also use the Yu-UYC with $b = 0.4$ as a linear approximation of the Huber-Mises yield criterion. The result obtained by Yu-UYC with $b = 0.4$ differ from the Huber-Mises yield criterion by 1.6% in the problem. This new criterion is much simpler and more convenient because of its linearity.

5   Some test results show that the value of b in carbon steel is generally between 0.5 and 1.0. We can adopt the criterion with the mean value of $b = 3/4$. In this case, the allowable maximum torque obtained by Yu-UST with $b = 3/4$ increased by 27.1% and 10.3%, respectively, compared with the Tresca yield criterion (Yu-UST with $b = 0$) and the Huber-Mises yield criterion $(b = (\sqrt{3} - 1)/2 = 0.366)$.

5.3   Thickness design of a thin-walled pressure vessel

The wall thickness of a thin-walled pressure vessel with known pressure and allowable stress can be obtained by using Yu-UST, as shown in Eq. 5.6.

$$t \geq \frac{2 + 2b - \alpha b}{1 + b} \frac{pD}{4[\sigma]} \tag{5.6}$$

For materials with the same tensile and compressive strength, Eq. 5.6 can be simplified as follows:

$$t \geq \frac{2 + b}{1 + b} \frac{pD}{4[\sigma]} \tag{5.7}$$

If the cylinder diameter of the thin-walled pressure tube is $D = 200cm$, the yield limit of the low-carbon steel material is $\sigma_y = 200MPa$, the safety factor is $n = 1.5$ and the internal pressure is $p = 1.6MPa$ The thickness of the wall t under the different strength theory can be calculated, as shown in Table 5.2. The relationship between the wall thickness of the thin-walled cylinder and the parameter b of the Yu-UST is shown in Figure 5.3.

Several conclusions can be obtained as follows:

1   The maximum difference of wall thickness between different strength theories is 25%.
2   The result obtained by the Yu-UYC with $b = 0$ is the same as the result obtained by the Tresca yield criterion.
3   The result of the Yu-UYC with $b = (\sqrt{3} - 1)/2 = 0.366$ is the same as the result obtained by the Huber-Mises yield criterion in this case.
4   The difference between the Yu-UYC with $b = 0.4$ and the Huber-Mises yield criterion is 1%. For convenience, the Yu-UYC with $b = 0.4$ can be used as a linear approximation of the Huber-Mises yield criterion.
5   The wall thickness of thin-walled cylinder for the materials with different tensile and compressive strength ratios ($\alpha < 1$) are shown in Figure 5.3.

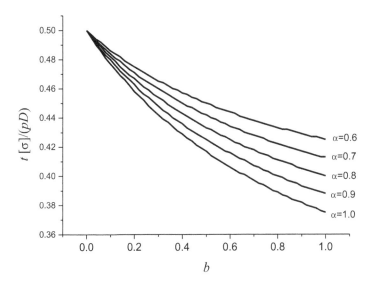

Figure 5.3 The relationship between wall thickness and the Yu-UST parameter b

Table 5.2 Calculation of the wall thicknesses of different strength theories

Strength theory	Wall thickness (mm)	Decreased thickness t compared with the Tresca yield criterion (%)	Decreased thickness t compared with the Huber-Mises yield criterion (%)
Yu-UST, $b = 0$	12	0	
Tresca yield criterion	12	0	
Yu-UST, $b = 1/4$	10.8	10.0	
Yu-UST, $b = (\sqrt{3} - 1)/2$	10.4	13.3	
Huber-Mises yield criterion	10.4	13.3	0
Yu-UST, $b = 0.4$	10.3	14.2	1
Yu-UST, $b = 1/2$	10	16.7	3.8
Yu-UST, $b = 3/4$	9.43	21.4	9.3
Yu-UST, $b = 1.0$	9.0	25	13.5

5.4   Strength design of a gun tube of trench mortar

The mortar is a widely used weapon in conventional wars, as shown in Figure 5.4. The barrel of mortar can be regarded as a thin-wall pressure cylinder. The results obtained in section 5.3 for the thickness design of a thin-wall pressure cylinder can be used for the thickness design of the mortar barrel.

Generally, the bore pressure is varying along the barrel axis. Consider that if a mortar barrel with diameter $D = 60$mm is subjected to internal pressure with a maximum $p = 60$MPa, the wall thickness of the barrel at the maximum pressure point can be obtained by using the Yu-UYC. The barrel is made of a steel with axial yield strength $\sigma_y = 650$MPa and shear yield strength $\tau_y = 420$MPa. The factor of safety is $n = 2.0$. The allowable yield stress can be obtained as $[\sigma_y] = \sigma_y/n = 325$MPa. The comparison among the wall thickness results obtained by different strength theories (i.e. Mariotte-St. Venant criterion, Tresca criterion, and Huber-Mises criterion under the maximum bore pressure $p = 60$MPa) is shown in Table 5.3.

As shown in Table 5.3, the obtained results of wall thickness are different by using different strength theories, and the shear strength is not considered in these single-strength theories (i.e. Tresca criterion,

Figure 5.4 Sketch of mortar

Table 5.3 Wall thicknesses obtained by different strength theories

Material strength parameter	Strength theory	Minimum wall thickness	Wall thickness (mm)
Axial yield strength: $\sigma_y = 650$MPa	Mariotte-St. Venant criterion	$t = \dfrac{2\text{-}\nu}{4} \cdot \dfrac{pD}{[\sigma_y]}$	4.71
Shear yield strength: $\tau_y = 420$MPa	Tresca criterion	$t = \dfrac{1}{2} \cdot \dfrac{pD}{[\sigma_y]}$	5.54
Poisson ratio: $\nu = 0.3$	Huber-Mises criterion	$t = \dfrac{\sqrt{3}}{4} \dfrac{pD}{[\sigma_y]}$	4.80

Mariotte-St. Venant criterion and Huber-Mises criterion). A more reasonable result can be obtained by using the Yu-UYC as follows:

The ratio of shear strength to axial tensile strength of the material is

$$b = \frac{2\tau_y - \sigma_y}{\sigma_y - \tau_y} = 0.83$$

The mathematic equation of Yu-UYC when $b = 0.83$ is

$$f = \sigma_1 - \frac{1}{1.83}(0.83\sigma_2 + \sigma_3) = \sigma_y, \text{ when } \sigma_2 \le \frac{\sigma_1 + \sigma_3}{2} \qquad (5.8a)$$

$$f' = \frac{1}{1.83}(\sigma_1 + 0.83\sigma_2) - \sigma_3 = \sigma_y, \text{ when } \sigma_2 \geq \frac{\sigma_1 + \sigma_3}{2} \qquad (5.8b)$$

According to Eq. 5.1, the stress in thin-walled pressure vessel satisfies $\sigma_2 = \frac{\sigma_1 + \sigma_3}{2}$, so either Eqs. 5.8a or 5.8b can be used. We choose Eq. 5.8a and obtain

$$f = \sigma_1 - \frac{1}{1.83}(0.83\sigma_2 + \sigma_3) = \frac{pD}{2t} - \frac{0.83}{1.83}\frac{pD}{4t} = \sigma_t \qquad (5.9)$$

The minimum wall thickness based on Yu-UYC is

$$t = \frac{2.83}{1.83}\frac{pD}{4[\sigma_y]} = 4.28\text{mm} \qquad (5.10)$$

The yield criterion obtained by the Yu-UST can describe the yield of material more reasonably than the single yield criterion because it is the combination of the theatrical basic and experimental results.

5.5   Strength design of an oil pipe

Pipeline systems are commonly used to transport crude oil, natural gas, sewage, water and other materials. The efficiency of pipelines is partly attributable to their practicality and safety. For a piping system to be used safely, it is necessary that some knowledge of the maximum pressure load it can support without leakage or catastrophic fracture be available to the designer and user. Consequently, accurate prediction of their burst pressure is an important consideration in the design for safety and integrity assessment of pipelines. The Yu-UYC can also be used to predict the burst pressure analysis of thin-walled oil pipes.

An example of the analysis of the oil pipe is given by Wang and Zhang (2011). The experimental result given by Zheng and Wen (2002) agreed with the result obtained by Wang and Zhang by using Yu-UYC with $b = 1$ (i.e. twin-shear strength theory). It can be found that the burst pressure can be increased by 43% compared with the result by using the Tresca yield criterion.

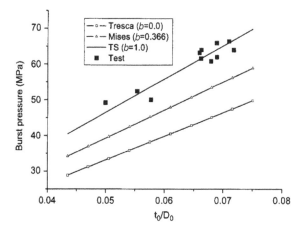

Figure 5.5 Comparison of analytical and experimental burst pressure for Q235A mild steel cylindrical chambers (Wang and Zhang, 2011)

5.6 Strength design of an automobile transmission shaft: maximum allowable torque

The main drive shaft of a vehicle's transmission power is made of seamless steel tube as shown in Figure 5.6. The inner diameter is d, the outer diameter is D, the wall thickness is t and the allowable tensile strength of this material is σ_y.

The expressions of the Yu-UYC are

$$f = \sigma_1 - \frac{1}{1+b}(b\sigma_2 + \sigma_3) = \sigma_y, \text{ when } \sigma_2 \leq \frac{\sigma_1 + \sigma_3}{1+\alpha} \tag{5.11a}$$

$$f' = \frac{1}{1+b}(\sigma_1 + b\sigma_2) - \sigma_3 = \sigma_y, \text{ when } \sigma_2 \geq \frac{\sigma_1 + \sigma_3}{1+\alpha} \tag{5.11b}$$

The stress condition equation of the main drive shaft satisfies both the first expression, Eq. 5.11a, and the second expression, Eq. 5.11b. Therefore, both the Yu-UYC expressions, Eq. 5.11a and 5.11b, can be selected.

If the inner diameter is $d = 85\text{mm}$, the outer diameter is $D = 90\text{mm}$, the wall thickness is $t = 2.5\text{mm}$ and the allowable tensile strength of this material is $\sigma = 120\text{MPa}$, the maximum allowable torque can be calculated as follows:

First, the internal and external diameter ratio of the main transmission shaft is $\alpha = d/D = 85/90 = 0.944$, and the torsion section modulus is

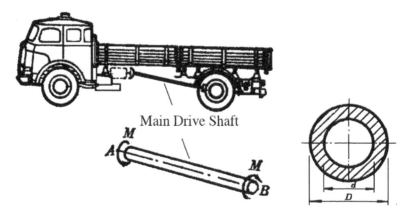

Figure 5.6 The force analysis of the main drive shaft of vehicle transmission power

$$W_p = \frac{\pi D^3}{16}(1-\alpha^4) = \frac{\pi \times 90^3}{16}(1-0.944^4)\text{mm}^3 = 29 \times 10^3 \text{mm}^3 = 29.5 \times 10^{-6}\text{m}^3.$$

The maximum shear stress of the axle is $\tau_{\max} = \dfrac{M_n}{W_p} = \dfrac{M_n}{29.5 \times 10^{-6}}$, and

the stress state of the main driving shaft at the dangerous point is

$$\sigma_1 = \tau, \quad \sigma_2 = 0, \quad \sigma_3 = -\tau, \text{ i.e. } |\sigma_1| = |\sigma_3| = |\tau| = \frac{M_n}{29.5 \times 10^{-6}}.$$

Second, we substitute the main drive shaft stress into Eq. 5.11a and get

$$\begin{aligned}
f &= \sigma_1 - \frac{1}{1+b}(b\sigma_2 + \sigma_3) = \tau - \frac{1}{1+b}(-\tau) = \frac{2+b}{1+b}\tau \\
&= \frac{2+b}{1+b} \times \frac{M_n}{29.5 \times 10^{-6}} = [\sigma] = 120 \times 10^6
\end{aligned} \tag{5.12}$$

Therefore, the maximum allowable torque is

$$M_n = 3540\left(\frac{1+b}{2+b}\right) \tag{5.13}$$

Taking the different values of the Yu-UYC parameter b, the maximum allowable torque of the transmission shaft under different strength theories can be obtained as shown in Table 5.4.

Table 5.4 The maximum allowable torque of the transmission shaft under different strength theories

Strength theory	Maximum allowable torque $[M_n]$ (Nm)	Pressure increases compared with the Tresca yield criterion (%)	Pressure increases compared with the Huber-Mises yield criterion (%)
Yu-UUC $b = 0$	1770	0	
Tresca yield criterion	1770	0	
Yu-UYC $b = 1/4$	1970	11.3	
Yu-UYC, $b = (\sqrt{3} - 1)/2$	2040	15.4	
Huber-Mises yield criterion	2040	15.4	0
Yu-UYC $b = 0.4$	2065	16.7	1.2
Yu-UYC $b = 1/2$	2120	19.8	3.9
Yu-UYC $b = 3/4$	2250	27.1	10.3
Yu-UYC $b = 1.0$	2360	33.3	15.7

We can draw several conclusions from the above results:

1   The results obtained by the Tresca yield criterion are the same as the results obtained by the Yu-UYC with $b = 0$, which means that the result obtained by the Tresca yield criterion is a special case of the series results of the Yu-UYC.

2   The results obtained by the Huber-Mises yield criterion are the same as the result obtained by the Yu-UYC with $b = 0.366$ in this case. The Yu-UYC with $b = 0.366$ can be seen as a linear approximation of the Huber0-Mises yield criterion.

3   The result obtained by the Yu-UST with $b = 0.4$ is almost the same as the result by the Huber-Mises yield criterion. The difference between them is 1%. Therefore, the Yu-UST with $b = 0.4$ can also be used as the linear approximation of the Huber-Mises yield criterion.

4   The greater allowable drive shaft torque increased with the growth of parameter b. The result obtained by Yu-UST with $b = 1.0$ can increase 33.3% compared with the result obtained by Tresca yield criterion.

5   Some experimental results show that the b value of carbon steel is generally between 0.5 and 1. An average value (i.e. $b = 3/4$) is used for the application. The allowable maximum torque by using

Yu-UST with $b = 3/4$ can be increased by 27.1% and 10.3%, respectively, compared with the Tresca yield criterion and the Huber-Mises yield criterion.

5.7   Strength design of an automobile transmission shaft: outer diameter design of drive shaft

If the internal and external diameter ratio of the automobile transmission shaft is constant and we choose $\alpha = d/D = 0.944$ as an example, the maximum torque of the automobile transmission shaft will be $M_n = 1770$ Nm. The outer diameter D and inner diameter d ($d/D = 0.944$) of the axle are designed according to different strength theories.

It can be obtained from the above Yu-UST equation that

$$\frac{M_n}{W_p}\left(\frac{2+b}{1+b}\right) = [\sigma] \tag{5.14}$$

$$W_p = \frac{\pi D^3}{16}(1 - \alpha^4) = \frac{M_n}{[\sigma]}\left(\frac{2+b}{1+b}\right) \tag{5.15}$$

$$D^3 = \frac{16}{\pi(1-\alpha^4)}\frac{M_n}{[\sigma]}\left(\frac{2+b}{1+b}\right) = \frac{16}{\pi(1-0.944^4)}\frac{1770}{120}\left(\frac{2+b}{1+b}\right) \tag{5.16}$$

$$= 365 \times 10^{-6}\left(\frac{2+b}{1+b}\right)$$

The comparisons and design results of the different theories at the same maximum torque are shown in Table 5.5.

Several conclusions can be obtained as follows:

1   The results obtained by Tresca yield criterion and Huber-Mises yield criterion can be deduced or linearly approximated by using the special case of the Yu-UYC, respectively.
2   The greater the number of b, the smaller cross-section area of the transmission shaft will be. It can be decreased by 17.4% compared with the Tresca yield criterion.
3   Some experimental results show that the b value of carbon steel is generally between 0.5 and 1. An average value of $b = 3/4$ is taken for the application. The cross-section area can be decreased by 14.8% and 6.2%, respectively, compared with the Tresca yield criterion and the Huber-Mises yield criterion.

Table 5.5 Comparisons of the different strength theories at the same maximum torque

Strength theory	Cross-section design of the transmission shaft			Pressure increases compared with the Tresca yield criterion (%)	Pressure increased compared with the Huber-Mises yield criterion (%)
	Outer diameter	Inner diameter	Cross section area		
Yu-UYC b = 0	90	85	692	0	
Tresca yield criterion	90	85	692	0	
Yu-UYC b = 1/4	86.9	82	646	6.77	
Yu-UYC b = ($\sqrt{3}$ − 1)/2	85.8	81	629	9.12	
Huber-Mises yield criterion	85.8	81	629	9.12	0
Yu-UYC b = 0.4	85.5	80.7	625	9.75	0.6
Yu-UYC b = 1/2	84.7	80	613	11.4	2.5
Yu-UYC b = 3/4	83.1	78.4	590	14.8	6.2
Yu-UYC b = 1.0	81.8	77.2	572	17.4	9.1

5.8 Strength design of a transmission axle

The automobile transmission shaft we described is only applied with torque. There are many transmission shafts in engineering. For these shafts, not only torque is applied, but blending moment is applied as well. The motor shaft has two moments, torque and blending, as shown in Figure 5.7. The material of the motor shaft will be subjected the σ-τ combined stresses. Sometimes the shaft is subjected to the torque moment and axial force. The material of the element will be acted under σ-τ combined stresses, as shown in Figure 5.8.

In Figure 5.8c, upper two figures, the normal stress is caused by axial force, and the shear stress is caused by torque moment. The combination of these two figures is equal to the lower two figures in Figure 5.8c.

In these two cases (Figure 5.7 and 5.8), the stress states are both the σ-τ combined stress state. Readers can also use the Yu-UYC as described in section 4.6 in this book for these two cases. A series of yield loci of Yu-UYC in the σ-τ stress state is shown in Figure 5.9.

The Yu-UYC is easy to use. While using the following Yu-UYC equations, we need to be careful to choose Eq. 5.17a or 5.17b, which depends on the stress state.

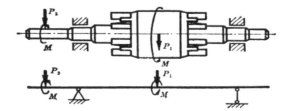

Figure 5.7 Motor shaft

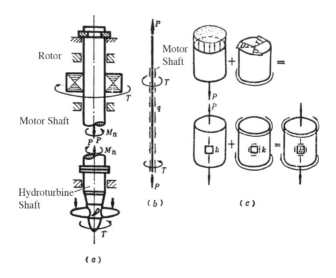

Figure 5.8 The rotor of a hydroturbine

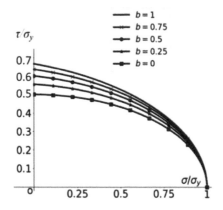

Figure 5.9 Yield loci of the Yu-UYC in the $\sigma-\tau$ stress state

$$f = \frac{2+b}{2+2b}\sqrt{\sigma^2 + 4\tau^2} + \frac{b}{2+2b}\sigma = \sigma_y, \text{ when } \sigma \geq 0 \qquad (5.17a)$$

$$f' = \frac{2+b}{2+2b}\sqrt{\sigma^2 + 4\tau^2} - \frac{b}{2+2b}\sigma = \sigma_Y, \text{ when } \sigma \leq 0 \qquad (5.17b)$$

5.9 Unified solution for elastic limit pressure of a thick-walled cylinder subjected to internal pressure

Thick-walled cylinders have been widely used in various projects. The study and calculation of elastic limit and plastic limit for a thick-walled cylinder subjected to internal pressure are of great significance both theoretically and practically. Much work has been done on the elastoplastic analysis of thick-walled cylinders subjected to internal pressure. In general, the Tresca yield criterion and the Huber-Mises yield criterion have been widely applied. However, the elastic and plastic pressure deduced from these two criteria can only be applied to a small number of materials. The Tresca yield criterion ignores the influence of intermediate principal stress on material yield, and its ultimate pressure calculation leads to conservative prediction. The Huber-Mises criterion is inconvenient because of its nonlinear mathematical expression.

The emergence of Yu-UST provides a new theoretical basis for elastoplastic analysis of thick-walled cylinders. Modern thick-walled cylinders often use high-strength steels, which have different tensile and compressive strengths. Therefore, we need to consider the SD effect (tensile strength and compressive strength difference) of materials, and the SD effect is also referred to as tensile strength asymmetry in material science. The two-parameter strength theory should be adopted at this time. We use the Yu-UST to study the unified solution of the elastic limit pressure and the plastic limit pressure of the thick-walled cylinder under internal pressure in a comprehensive description in sections 5.6 and 5.7.

A thick-walled cylinder under internal pressure p with inner radius r_a and outer radius r_b, respectively, is shown in Figure 5.10. It is assumed that the thick-walled cylinder has a certain length and remains the same deformation in its middle section. The cross section of the thick-walled cylinder remains flat during the force expansion process. This means that the longitudinal strain ε_z is independent of the radius.

On the section far from the ends of the thick-walled cylinder, its stress, strain and equilibrium equation is

$$\frac{d\sigma_r}{dr} = \frac{\sigma_\theta - \sigma_r}{r} \qquad (5.18)$$

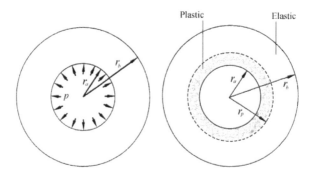

Figure 5.10 A thick-walled cylinder under the internal pressure p

where z-axis is the longitudinal axis of the thick-walled cylinder and the cylindrical coordinate system is (r, θ, z). According to the generalized Hooke's law, the longitudinal stress of a thick-walled cylinder at elastic stage is

$$\sigma_z = E\varepsilon_z + \nu(\sigma_r + \sigma_\theta) \tag{5.19}$$

where E is the elastic modulus and ν is the Poisson ratio. Radial strain ε_r and circumferential strain ε_θ equal:

$$\varepsilon_r = -\nu\varepsilon_z + \frac{1+\nu}{E}\left[(1-\nu)\sigma_r - \nu\sigma_\theta\right] \tag{5.20a}$$

$$\varepsilon_\theta = -\nu\varepsilon_z + \frac{1+\nu}{E}\left[(1-\nu)\sigma_\theta - \nu\sigma_r\right] \tag{5.20b}$$

The harmonic equation $\dfrac{d}{dr}(\sigma_r + \sigma_\theta) = 0$ shows that $\sigma_r + \sigma_\theta$ remains constant. Integrating Eq. 5.18 and considering the boundary conditions $\sigma_r = 0$ $(r = r_b)$ and $\sigma_r = -p$ $(r = r_a)$, we can get the stress equation as follows:

$$\sigma_r = -p\left(\frac{\dfrac{r_b^2}{r^2} - 1}{\dfrac{r_b^2}{r_a^2} - 1}\right), \quad \sigma_\theta = p\left(\frac{\dfrac{r_b^2}{r^2} + 1}{\dfrac{r_b^2}{r_a^2} - 1}\right) \tag{5.21}$$

This is the solution that Lamé obtained in 1852. If the thick-walled cylinder has a inner pressure P, the axial stress $\sigma_z = P/\pi(r_b^2 - r_a^2)$ will act uniformly on the cross section. $P = 0$ indicates that the thick-walled cylinder is open; the $P = \pi r_a^2 p$ indicates that the thick-walled cylinder is closed. For plane strain problems ($\varepsilon_z = 0$), the stress σ_z can be directly obtained from Eqs. 5.19 and 5.20. The stress formulas under the three conditions are, respectively, as follows:

$$\sigma_z = \frac{p}{K^2 - 1} \quad \text{Closed} \tag{5.22a}$$

$$\sigma_z = 0 \quad \text{Opened} \tag{5.22b}$$

$$\sigma_z = \frac{2\nu p}{K^2 - 1} \quad \text{Plane strain} \tag{5.22c}$$

The corresponding axial strain is

$$\varepsilon_z = \frac{(1 - 2\nu)p}{(K^2 - 1)E} \quad \text{Closed} \tag{5.23a}$$

$$\varepsilon_z = 0 \quad \text{Opened} \tag{5.23b}$$

$$\varepsilon_z = \frac{-2\nu p}{(K^2 - 1)E} \quad \text{Plane strain} \tag{5.23c}$$

Under these three conditions, σ_z is the intermediate principal stress. For the closed-end case, σ_z is equal to the average value of the other two principal stresses. If the material is incompressible in the elastic and plastic stages, the longitudinal stress σ_z under the plane strain condition is the same as the closed-end condition. They are $\sigma_1 = s_\theta$, $\sigma_2 = \sigma_z$, $\sigma_3 = s_r$, and

$$\sigma_2 = \frac{1}{2}(\sigma_1 + \sigma_3) \le \frac{\sigma_1 + \alpha\sigma_3}{1 + \alpha} \tag{5.24}$$

This condition satisfied the first expression of the Yu-UST.

$$\sigma_1 - \frac{\alpha}{1 + b}(b\sigma_2 + \sigma_3) = \sigma_t \tag{5.25}$$

The yield conditions of thick-walled cylinders at closed ends and plane strain conditions can be expressed as:

$$\frac{2 + (2 - \alpha)\alpha}{2(1 + b)}\sigma_\theta - \frac{\alpha(2 + b)}{2(1 + b)}\sigma_r = \sigma_t \tag{5.26}$$

The yield conditions of thick-walled cylinders at opened ends:

$$\sigma_\theta - \frac{\alpha}{1+b}\sigma_r = \sigma_t \tag{5.27}$$

Substituting Eq. 5.21 into Eqs. 5.26 and 5.27, we get:

$$[2 + (2-\alpha)b]\frac{p}{K^2-1}\left(\frac{r_b^2}{r^2}+1\right) + \alpha(2+b)\frac{p}{K^2-1}\left(\frac{r_b^2}{r^2}-1\right) \tag{5.28}$$
$$= 2(1+b)\sigma_t$$

The elastic limit pressure of a thick-walled cylinder under Yu-UST is obtained as follows:

$$p_e = \frac{(1+b)(K^2-1)\sigma_t}{K^2(1+b+\alpha)+(1+b)(1-\alpha)} \quad \text{Closed} \tag{5.29}$$

$$p_e = \frac{(1+b)(K^2-1)\sigma_t}{(1+b)(K^2+1)+\alpha(K^2-1)} \quad \text{Opened} \tag{5.30}$$

$$p_e = \frac{(1+b)(K^2-1)\sigma_t}{K^2(1+b+\alpha)+(1+b)(1-\alpha)} \quad \text{Plane strain} \tag{5.31}$$

The limit pressure of the closed end thick-walled cylinder under the Mohr-Coulomb condition is a special case when the unified solution is at $b = 0$.

$$p_e = \frac{K^2-1}{(1+\alpha)K^2+(1-\alpha)}\sigma_t \quad \text{Single-shear theory} \tag{5.32}$$

The limit pressure of the closed end thick-walled cylinder under the twin-shear theory condition is a special case when the unified solution is at $b = 1$.

$$p_e = \frac{2(K^2-1)}{(2+\alpha)K^2+2(1-\alpha)}\sigma_t, \quad \text{Twin-shear theory} \tag{5-33}$$

Materials with the same tensile strength and compressive strength (i.e. $\sigma_t = \sigma_c = \sigma_y$, Eq. 5.29, 5.30 and 5.31) can be simplified as:

$$p_e = \frac{(1+b)(K^2-1)}{K^2(2+b)}\sigma_y \quad \text{Closed} \tag{5.34}$$

$$p_e = \frac{(1+b)(K^2-1)}{K^2(2+b)+b}\sigma_y \quad \text{Opened} \tag{5.35}$$

$$p_e = \frac{(1+b)(K^2 - 1)}{K^2(2+b) + b(1-2\nu)}\sigma_y \quad \text{Plane strain} \tag{5.36}$$

This result is the same as that obtained by the Yu-UYC of Wang and Fan in 1998. The elastic limit pressure obtained from the Tresca yield criterion can be obtained from the condition of $\alpha = 1$ and $b = 0$. The results are the same as those obtained by Eqs. 5.34, 5.35 and 5.36. It can be seen that the Tresca yield criterion does not consider the intermediate principal stress. Therefore, it cannot reflect three different situations: the closed end, the opened end and the plane strain.

$$p_e = \frac{K^2 - 1}{2K^2}\sigma_y \tag{5.37}$$

For the Huber-Mises yield criterion, the elastic limit pressure can be approximated by the Yu-UYC under the condition of $\alpha = 1$ and $b = 1/2$.

$$p_e = \frac{3(K^2 - 1)}{5K^2}\sigma_y \quad \text{Closed} \tag{5.38}$$

$$p_e = \frac{3(K^2 - 1)}{5K^2 + (1 - 2\nu)}\sigma_y \quad \text{Plane strain} \tag{5.39}$$

The classical Huber-Mises yield criterion materials are solved as

$$p_e = \frac{K^2 - 1}{\sqrt{3K^2}}\sigma_y \quad \text{Closed} \tag{5.40}$$

$$p_e = \frac{K^2 - 1}{\sqrt{3K^4 + 1}}\sigma_y \quad \text{Opened} \tag{5.41}$$

$$p_e = \frac{K^2 - 1}{\sqrt{3K^4 + (1 - 2\nu)^2}}\sigma_y \quad \text{Plane strain} \tag{5.42}$$

The difference between the results obtained from the Yu-UYC under the condition of $\alpha = 1$ and $b = 1/2$ and the classical solution is only 0.38%.

The elastic limit pressure of the twin-shear yield criterion can be obtained from the Yu-UYC under the condition $\alpha = 1$ and $b = 1$.

$$p_e = \frac{2(K^2 - 1)}{3K^2}\sigma_y \quad \text{Closed} \tag{5.43}$$

$$p_e = \frac{2(K^2 - 1)}{3K^2 + 1}\sigma_y \quad \text{Opened} \tag{5.44}$$

$$p_e = \frac{2(K^2 - 1)}{3K^2 + (1 - 2\nu)} \sigma_y \quad \text{Plane strain} \tag{5.45}$$

The difference of the elastic limit pressure of the twin-shear yield criterion and single-shear yield criterion (Tresca yield criterion) is 33.4%.

We can see that all kinds of solutions in the past can be obtained from the unified solution obtained by the Yu-UST with different parameters α and b. Figure 5.11 shows the relationship between the ultimate pressure of a thick-walled cylinder with the ratio of the inner and outer diameter $k = 2$ and the Yu-UST parameter α and b. The application of the Yu-UST will increase the calculation limit pressure of thick-walled cylinders. It will also achieve greater economic benefits for suitable materials.

The equations for calculating the elastic pressure of the thick-walled cylinder and the elastic pressure at the opened end of the thick-walled cylinder are summarized in Tables 5.6 and 5.7, respectively.

When the thick-walled cylinder is subjected to uniform external pressure P, it can be obtained by similar analysis.

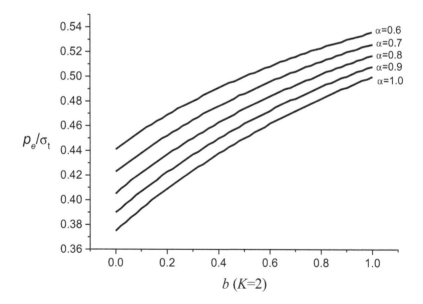

Figure 5.11 Relationship between elastic limit pressure of a thick-walled cylinder and Yu-UST parameter b

Table 5.6 Equations for calculating the elastic pressure of a thick-walled cylinder

	Materials		The elastic limit pressure
1	The unified solution of the elastic limit pressure for non-SD material $$p_e = \frac{(1+b)(K^2-1)}{K^2(2+b)}\sigma_y$$		
2	Materials with $\alpha = 1$	$b = 0$	$$p_e = \frac{K^2-1}{2K^2}\sigma_y$$
3		Huber-Mises materials	$$p_e = \frac{K^2-1}{\sqrt{3K^2}}\sigma_y$$
4		$b = 1/2$	$$p_e = \frac{3(K^2-1)}{5K^2}\sigma_y$$
5		$b = 1$	$$p_e = \frac{2(K^2-1)}{3K^2}\sigma_y$$
6	The unified solution of the elastic limit pressure for SD material $$p_e = \frac{(1+b)(K^2-1)\sigma_t}{(K^2+1)(1+b)+\alpha(K^2-1-b)}$$		
7	Materials with $\alpha \neq 1$	$b = 0$	$$p_e = \frac{K^2-1}{(1+\alpha)K^2+(1-\alpha)}\sigma_t$$
8		$b = 1/2$	$$p_e = \frac{3(K^2-1)\sigma_t}{3(K^2+1)+2\alpha(K^2-1.5)}$$
9		$b = 1$	$$p_e = \frac{2(K^2-1)}{K^2(2+\alpha)+2(1-\alpha)}\sigma_t$$

Table 5.7 Equations for calculating the elastic pressure at the opened end of a thick-walled cylinder

	Materials		The elastic limit pressure
1	The unified solution of the elastic limit pressure for non-SD material $$p_e = \frac{(1+b)(K^2-1)}{K^2(2+b)+b}\sigma_y$$		
2	Materials with $\alpha = 1$	$b = 0$	$$p_e = \frac{K^2-1}{2K^2}\sigma_y$$
3		Huber-Mises materials	$$p_e = \frac{K^2-1}{\sqrt{3K^4+1}}\sigma_y$$

	Materials		The elastic limit pressure
4		$b = 1/2$	$p_e = \dfrac{3(K^2 - 1)}{5K^2 + 1}\sigma_y$
5		$b = 1$	$p_e = \dfrac{2(K^2 - 1)}{3K^2 + 1}\sigma_y$
6	The unified solution of the elastic limit pressure for SD material $\dfrac{(1+b)(K^2 - 1)\sigma_t}{(1+b)(K^2 + 1) + \alpha(K^2 - 1)}$		
7	Materials with $\alpha \neq 1$	$b = 0$	$p_e = \dfrac{K^2 - 1}{(1+\alpha)K^2 + (1-\alpha)}\sigma_t$
8		$b = 1/2$	$p_e = \dfrac{3(K^2 - 1)}{3(1 + K^2) + 2\alpha(K^2 - 1)}\sigma_t$
9		$b = 1$	$p_e = \dfrac{2(K^2 - 1)}{2(K^2 + 1) + \alpha(K^2 - 1)}\sigma_t$

5.10   Analysis of plastic ultimate pressure for a thick-walled cylinder with internal pressure

When the internal pressure of a thick-walled cylinder exceeds the elastic limit pressure of p_e, plastic zone will begin to appear in the inner wall of thick-walled cylinder and expand outward. r_c is the radius of the elastoplastic outer boundary of the thick-walled cylinder. The application of boundary conditions for radial and circumferential stresses ($\sigma_r = 0, r = r_b$) in the elastic range, $r_c \leq r \leq r_b$, can be obtained by the Lamé equation, and the yield conditions are satisfied at $r = r_c$. When the plastic zone reaches the outer surface of the thick-walled cylinder, the pressure of the thick-walled cylinder reaches the maximum value, i.e. the plastic limit pressure.

In the elasto-plastic stage of thick-walled cylinder, the elastic part can be regarded as a new thick-walled cylinder with inner radius r_c, outer radius r_b and internal pressure p_e. The stresses in the elastic region are

$$\sigma_\theta = \frac{p_e r_c^2}{r_b^2 - r_c^2}\left(1 + \frac{r_b^2}{r^2}\right) \tag{5.46}$$

$$\sigma_r = \frac{p_e r_c^2}{r_b^2 - r_c^2}\left(1 - \frac{r_b^2}{r^2}\right) \tag{5.47}$$

$$\sigma_z = \frac{\nu}{2}(\sigma_\theta + \sigma_r) \tag{5.48}$$

in these equations,

$$p_e = \frac{2(1+b)(r_b^2 - r_c^2)}{(2+2b-\alpha b)(r_b^2 + r_c^2) + \alpha(2+b)(r_b^2 - r_c^2)} \sigma_t \tag{5.49}$$

In plastic region, the material produced plastic yielding, which satisfies the expression of the Yu-UST, Eq. 5.2a or 5.2b. The stress of the thick-walled cylinder satisfies $\sigma_2 = \frac{1}{2}(\sigma_1 + \sigma_3) \le \frac{\sigma_1 + \alpha\sigma_3}{1+\alpha}$. Therefore, the Yu-UST Eq. 5.2a is used as the yield criterion.

$$\frac{2+(2-\alpha)b}{2(1+b)}\sigma_\theta - \frac{\alpha(2+b)}{2(1+b)}\sigma_r = \sigma_t \tag{5.50}$$

Substituting Eq. 5.50 into the equilibrium Eq. 5.18

$$\frac{d\sigma_r}{dr} + \frac{2(1+b)(1-\alpha)}{2+(2-\alpha)b}\frac{\sigma_r}{r} - \frac{2(1+b)}{2+(2-\alpha)b}\frac{\sigma_t}{r} = 0 \tag{5.51}$$

The general solution of the differential equation is

$$\sigma_r = \frac{c}{r^{\frac{2(1+b)(1-\alpha)}{2+(2-\alpha)b}}} + \frac{\sigma_t}{1-\alpha} \tag{5.52}$$

According to the condition, $r = r_a$, $\sigma_r = -p$, Eq. 5.52 can be rewritten as

$$-p = \frac{c}{r_a^{\frac{2(1+b)(1-\alpha)}{2+(2-\alpha)b}}} + \frac{\sigma_t}{1-\alpha} \tag{5.53}$$

The integral constant is determined by the boundary condition.

$$c = (-p - \frac{\sigma_t}{1-\alpha})A^{\frac{2(1+b)(1-\alpha)}{2+(2-\alpha)b}} \tag{5.54}$$

Stresses in the plastic region ($r_a \le r \le r_c$) are shown as follows:

$$\sigma_r = -(p + \frac{\sigma_t}{1-\alpha})(\frac{r_a}{r})^{\frac{2(1+b)(1-\alpha)}{2+(2-\alpha)b}} + \frac{\sigma_t}{1-\alpha} \tag{5.55}$$

$$\sigma_\theta = \frac{2(1+b)\sigma_t}{2+(2-\alpha)b} - \frac{\alpha(2+b)}{2+(2-\alpha)b}$$

$$\left[\left(p+\frac{\sigma_t}{1-\alpha}\right)\left(\frac{r_a}{r}\right)^{\frac{2(1+b)(1-\alpha)}{2+(2-\alpha)b}} + \frac{\sigma_t}{1-\alpha}\right] \tag{5.56}$$

$$\sigma_z = \frac{1}{2}(\sigma_r + \sigma_\theta) \tag{5.57}$$

Equations 5.55, 5.56 and 5.57 give the stress formula for the plastic zone of a thick-walled cylinder.

For a certain pressure P, the radius r_c of plastic zone can be obtained by Eq. 5.55. When the pressure increases, the plastic zone radius r_c gradually expands from r_a to r_b. The continuity of the radial stress σ_r at $r = r_c$ is:

$$\sigma_{r=r_c} \text{ (elastic region)} = \sigma_{r=r_c} \text{ (plastic region)}$$

Substituting the equation of the radial stress, Eq. 5.47 and 5.55, into the stress continuity condition, the relation between pressure P and radius of plastic region can be obtained as Eq. 5.58. Therefore, the relationship between the radius of plastic region and the radius of the thick-walled cylinder with different sizes and materials can be calculated.

$$p = \left(\frac{r_c}{r_a}\right)^{\frac{2(1+b)(1-\alpha)}{2+(2-\alpha)b}} \left[\frac{2(1+b)(r_b^2 - r_c^2)}{(2+2b-\alpha b)(r_b^2 + r_c^2)+\alpha(2+b)(r_b^2 - r_c^2)} + \frac{1}{1-\alpha}\right]\sigma_t - \frac{\sigma_t}{1-\alpha} \tag{5.58}$$

According to Eq. 5.58, when the thick-walled cylinder enters the elastic plastic state from elastic state, the radius of plastic zone expands continuously. The formula obtained from Yu-UST also shows that the radius of plastic zone is related to the strength theory adopted. We calculate a thick-walled cylinder with five typical parameters $b = 0$, $b = 1/4$, $b = 1/2$, $b = 3/4$ and $b = 1$ of the Yu-UYC, and the plastic region expansion of the thick-walled cylinder under the same pressure is shown in Figure 5.12.

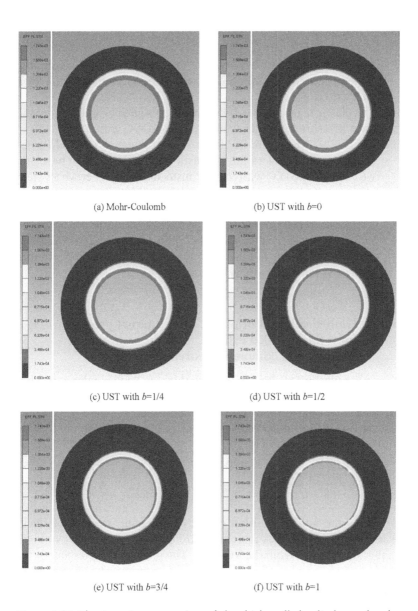

(a) Mohr-Coulomb

(b) UST with $b=0$

(c) UST with $b=1/4$

(d) UST with $b=1/2$

(e) UST with $b=3/4$

(f) UST with $b=1$

Figure 5.12 Plastic region expansion of the thick-walled cylinder under the same pressure

As we can see in Figure 5.12, the red color is the largest plastic deformation region of the thick-walled cylinder under the same pressure. Obviously:

1   The plastic region of $b = 0$ of the thick-walled cylinder under the same pressure is the largest. It is the same as the result of Tresca yield criterion.
2   The plastic region of $b = 1/4$ of the thick-walled cylinder under the same pressure is smaller than that of Yu-UYC with $b = 0$.
3   For $b = 1/2$, the radius of plastic zone of Huber-Mises yield criterion is similar to the radius of plastic zone of Yu-UYC. Therefore, we can use Yu-UYC with $b = 1/2$ instead of the Huber-Mises yield criterion. Yu-UYC with $b = 1/2$ is also the linear approximation of Huber-Mises yield criterion.
4   The plastic region of $b = 3/4$ of the thick-walled cylinder under the same pressure is smaller than that of Yu-UYC with $b = 1/2$.
5   The plastic region radius of $b = 1$ of the thick-walled cylinder under the same pressure is the smallest.

5.11   Unified solution of plastic limit internal pressure for a thick-walled cylinder with the same tensile strength and compressive strength materials

When the plastic region radius of the thick-walled cylinder r_c reaches the outer diameter r_b, the thick-walled cylinder is in the complete plastic state. The unified solution of the plastic limit pressure equation of the thick-walled cylinder with the same yield strength both in tension and compression, $\alpha = 1$, is

$$p_P = \frac{2(1+b)\sigma_t}{2+b} \ln K \tag{5.59}$$

This is the unified solution of the plastic limit pressure of thick-walled cylinder with the same tensile strength and compressive strength materials obtained from the Yu-UYC. The plastic limit pressure of the thick-walled cylinder of the Tresca yield criterion, the Huber-Mises yield criterion and the twin-shear yield criterion with the same tensile strength and compressive strength materials can be obtained as follows:

$$p_p = \sigma_t \ln K, \, b = 0, \text{ Tresca yield criterion} \tag{5.60}$$

$$p_P = \frac{6}{5}\sigma_t \ln K, \, b = 1/2, \text{ Huber-Mises yield criterion} \tag{5.61}$$

$$p_p = \frac{4}{3}\sigma_t \ln K, \, b = 1, \text{ Twin shear yield criterion} \tag{5.62}$$

5.12 Unified solution of plastic limit internal pressure for a thick-walled cylinder with different tensile strength and compressive strength materials

When the plastic region radius of the thick-walled cylinder r_c reaches the outer diameter r_b, the thick-walled cylinder is in the complete plastic state. The unified solution of the plastic limit pressure equation of the thick-walled cylinder with different strength and compressive strength, $\alpha \neq 1$, is

$$p_p = \frac{\sigma_t}{1-\alpha}\left(K^{\frac{2(1+b)(1-\alpha)}{2+2b-\alpha b}} - 1\right) \tag{5.63}$$

The unified solution in Eq. 5.58 can be applied to the thick-walled cylinder with a closed-end and plane strain state. The limit pressure solution of Mohr-Coulomb theory when $b = 0$ is obtained as follows:

$$p_p = \frac{\sigma_t}{1-\alpha}(K^{(1-\alpha)} - 1) \tag{5.64}$$

According to Eq. 5.64, the limit pressure solution of a thick-walled cylinder of twin-shear strength theory when $b = 1$ is obtained as follows:

$$p_p = \frac{\sigma_t}{1-\alpha}(K^{\frac{4(1-\alpha)}{4-\alpha}} - 1) \tag{5.65}$$

The relationship between the plastic limit pressure of the thick-walled cylinder with different radius ratio ($K = 1.8$, $K = 2.0$, $K = 2.5$, $K = 3.0$) and the parameter b of the Yu-UST is shown in Figures 5.14 to 5.17. The equations for calculating the plastic limit pressure at the closed end of the thick-walled cylinder are summarized in Table 5.8.

From these results, we can see that the strength theory has great influence on the calculation results. The application of Yu-UST can significantly increase the elastic limit pressure and plastic limit pressure of a thick-walled cylinder.

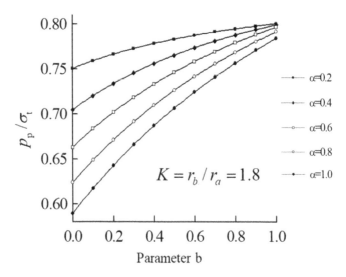

Figure 5.13 Relationship between the plastic limit pressure of a thick-walled cylinder with different radius ratios and the parameter b of the Yu-UST ($K = r_b/r_a = 1.8$)

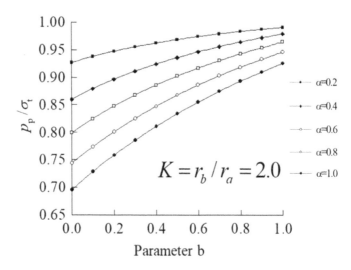

Figure 5.14 Relationship between the plastic limit pressure of a thick-walled cylinder with different radius ratios and the parameter b of the Yu-UST ($K = r_b/r_a = 2$)

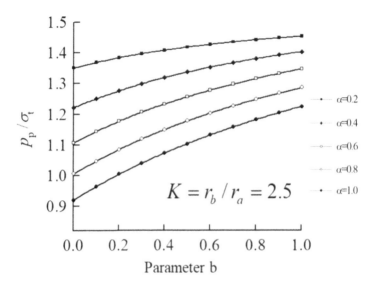

Figure 5.15 Relationship between the plastic limit pressure of a thick-walled cylinder with different radius ratios and the parameter b of the Yu-UST ($K = r_b/r_a = 2.5$)

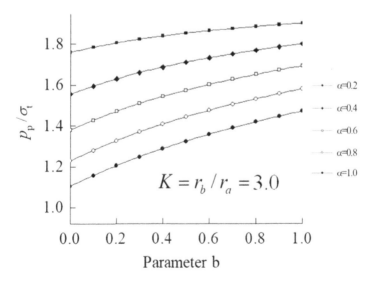

Figure 5.16 Relationship between the plastic limit pressure of the thick-walled cylinder with different radius ratios and the parameter b of the Yu-UST ($K = r_b/r_a = 3$)

Table 5.8 Equations for calculating the plastic limit pressure at the closed end of a thick-walled cylinder

	Materials		The plastic limit pressure
1	The unified solution of the plastic limit pressure for non-SD material $p_p = \dfrac{2(1+b)\sigma_t}{2+b}\ln K$		
2	Materials with $\alpha = 1$	$b = 0$	$p_p = \sigma_t \ln K$
3		$b = 0.4$	$p_p = \dfrac{2.8\sigma_t}{2.4}\ln K$
4		$b = 1/2$	$p_p = \dfrac{6}{5}\sigma_t \ln K$
5		$b = 1$	$p_p = \dfrac{4}{3}\sigma_t \ln K$
6	The unified solution of the plastic limit pressure for SD material $p_p = \dfrac{\sigma_t}{1-\alpha}(K^{\frac{2(1+b)(1-\alpha)}{2+2b-\alpha b}} - 1)$		
7	Materials with $\alpha \neq 1$	$b = 0$	$p_p = \dfrac{\sigma_t}{1-\alpha}(K^{(1-\alpha)} - 1)$
8		$b = 1/2$	$p_p = \dfrac{\sigma_t}{1-\alpha}(K^{\frac{6(1-\alpha)}{6-\alpha}} - 1)$
9		$b = 1$	$p_p = \dfrac{\sigma_t}{1-\alpha}(K^{\frac{4(1-\alpha)}{4-\alpha}} - 1)$

5.13   Barrel strength design of artillery

Figure 5.17 is a sketch of an artillery. The barrel of the artillery can be regarded as a thick-walled cylinder under internal pressure. Because the materials are under a complex stress state, the strength theory is needed for the strength design of an artillery barrel.

As we know, the limit pressure of a thick-walled cylinder is a function of different radius ratios. Consider that the barrel of an artillery is made of steel with axial yield strength $\sigma_y = 1250\text{MPa}$ and shear yield strength $\tau_y = 750\text{MPa}$. The factor of safety is $n = 1.5$. The allowable yield strength can be obtained as $[\sigma_y] = \dfrac{\sigma_y}{n} = 833.33\text{MPa}$. The maximum limit pressure of a thick-walled cylinder can be obtained by using different strength theories. The comparison among the results

Figure 5.17 Sketch of an artillery

Table 5.9 Limit pressure of the barrel of a artillery

Material strength parameter	Strength theory	Limit pressure (MPa)	Radius ratio (K)	Limit pressure (MPa)
Axial yield strength: $\sigma_y = 1250\text{MPa}$	Mariotte-St. Venant criterion	$P_p = \dfrac{3}{2}\sigma_p \dfrac{K^2-1}{2K^2+1}$	$K = 1.8$	374.33
Shear yield strength: $\tau_y = 680\text{MPa}$	Tresca criterion	$P_p = \sigma_p \dfrac{K^2-1}{2K^2}$		288.07
Poisson ratio: $\nu = 0.3$	Huber-Mises criterion	$P_p = \sigma_p \dfrac{K^2-1}{\sqrt{3K^4+1}}$		327.48

obtained by different strength theories (i.e. Mariotte-St. Venant criterion, Tresca criterion and Huber-Mises criterion) under the radius ratio $K = 1.8$ and the factor of safety $n = 1.5$ is shown in Table 5.9.

It is interesting to find that the Huber-Mises yield criterion is used in the design of artillery barrel in the United States, the Tresca criterion is used in Germany and the maximum tensile strain criterion is used in Russia and China. Which one of them is reasonable for the design of barrel? Can a better strength theory be used in this problem? How should a reasonable strength theory be chosen? Unified strength theory provides a new method to design the strength of a barrel and can take the place of the traditional yield criterion.

In this problem, the ratio of shear strength and axial tensile strength is

$$b = \frac{2\tau_y - \sigma_y}{\sigma_y - \tau_y} = 0.60 \tag{5.66}$$

The mathematic equation of Yu-UYC when $b = 0.60$ is

$$f = \sigma_1 - \frac{1}{1.60}(0.60\sigma_2 + \sigma_3) = \sigma_y, \text{ when } \sigma_2 \le \frac{\sigma_1 + \sigma_3}{2} \tag{5.67a}$$

$$f' = \frac{1}{1.60}(\sigma_1 + 0.60\sigma_2) - \sigma_3 = \sigma_y, \text{ when } \sigma_2 \ge \frac{\sigma_1 + \sigma_3}{2} \tag{5.67b}$$

According to Eq. 5.18 and Eq. 5.19, the stress in thick-walled pressure satisfies $\sigma_2 \le \dfrac{\sigma_1 + \sigma_3}{2}$, so we can choose Eq. 5.8a and obtain

$$p_e = \frac{(1+b)(K^2-1)}{K^2(2+b)+b}\sigma_y = \frac{1.60(K^2-1)}{2.60K^2+0.60}\sigma_y \tag{5.68}$$

The limit pressure based on Yu-UYC is

$$p = \frac{1.60(K^2-1)}{2.60K^2+0.60}\left[\sigma_y\right] = 330.97\text{MPa} \tag{5.69}$$

As we can see in the last chapter, the Yu-UYC can describe the yield of material more precisely than the single failure theory. So the yield criterion Eq. 5.67 (i.e. a special case of Yu-UYC) gives a more reasonable result than the single criterion.

5.14   The economic significance of the use of Yu-unified strength theory

For the design of the ultimate pressure of thick-walled cylinder, the elastic limit pressure and plastic limit pressure are the two important parameters. The unified solutions of elastic limit pressure and plastic limit pressure of thick-walled cylinders are obtained by using Yu-UST.

From the results in this chapter, we can see that the solutions obtained by the Yu-UST are a series of ordered solutions, covering all solutions from the single-shear theory to the twin-shear theory. The limit pressure of the traditional Tresca yield criterion can be degenerated by the unified solution of $b = 0$ and $\alpha = 1$; the ultimate pressure of the Huber-Mises yield criterion can be degenerated by the unified solution of $b = 0.4$ and $\alpha = 1$.

The results also show that the higher limit pressure can be obtained by using the unified solution of $b > 0$. The unified solution of $b = 1$ will increase the limit pressure by 33.4% compared with the unified solution of $b = 0$. The unified solution of $b = 1$ will increase the limit pressure by 15.5% compared with the Huber-Mises criterion. This may save a large amount of material. The application of Yu-UST provides a series of theoretical foundations for a more economical design.

Others, such as thick-walled cylinder bearing external pressure; thick-walled cylinder bearing internal and external pressures; and thick-walled cylinder bearing internal pressure, external pressure and torsion, can be studied by the same method.

5.15 Procedure of the applications of the Yu-UST and Yu-UYC

The expression for the Yu-UST is

$$F = \sigma_1 - \frac{\alpha}{1+b}(b\sigma_2 + \sigma_3) = \sigma_t, \text{ when } \sigma_2 \leq \frac{\sigma_1 + \alpha\sigma_3}{1+\alpha} \quad (5.70a)$$

$$F' = \frac{1}{1+b}(\sigma_1 + b\sigma_2) - \alpha\sigma_3 = \sigma_t, \text{ when } \sigma_2 \geq \frac{\sigma_1 + \alpha\sigma_3}{1+\alpha} \quad (5.70b)$$

The procedure of the applications of the Yu-UST and Yu-UYC can be stated as follows:

1 Determining the maximum stress point of the components and structures
2 Calculating the principal stresses σ_1, σ_2 and σ_3
3 The Yu-UST can be applied to materials with different tensile strength and compressive strength. The Yu-UYC can be applied to materials with the same yield strength both in tension and compression.

We must be careful to use the equations of the Yu-UST according to the stress state and additional conditions. If $\sigma_2 \leq \dfrac{\sigma_1 + \alpha\sigma_3}{1+\alpha}$, we should adopt the Eq. 5.70a; if $\sigma_2 \geq \dfrac{\sigma_1 + \alpha\sigma_3}{1+\alpha}$, we should use the Eq. 5.70b.

4 A certain parameter b can be selected for more accurate calculation. The Yu-UYC parameter b can be determined by the shear strength

τ_y of the materials with $\alpha = \sigma_t/\sigma_c = 1$ as shown in section 4.4. The relationship between them is

$$b = \frac{2\tau_y - \sigma_y}{\sigma_y - \tau_y} \tag{5.71}$$

Some experimental results shown in Table 4.1 can be used as the value of parameter b. It can be directly substituted into the Yu-UYC. The Tresca yield criterion can be applied to material with shear strength $\tau_y = 0.5\sigma_y$. The Huber-Mises yield criterion can be applied to material with shear strength $\tau_y = 0.577\sigma_y$.

5   Examples:

When $\tau_y = 0.5\sigma_y$, $b = 0$, this is the Tresca yield criterion.

When $\tau_y = 1.4/2.4\sigma_y$, we obtained parameter $b = 0.4$. This is the linear approximation of the Huber-Mises yield criterion. The equations are

$$f = \sigma_1 - \frac{1}{1.4}(0.4\sigma_2 + \sigma_3) = \sigma_y, \text{ when } \sigma_2 \le \frac{1}{2}(\sigma_1 + \sigma_3) \tag{5.72a}$$

$$f' = \frac{1}{1.4}(\sigma_1 + 0.4\sigma_2) - \sigma_3 = \sigma_y, \text{ when } \sigma_2 \ge \frac{1}{2}(\sigma_1 + \sigma_3) \tag{5.72b}$$

When $\tau_y = 2/3\sigma_y$, we obtained parameter $b = 1$. This is the twin-shear yield criterion. The equations are

$$f = \sigma_1 - \frac{1}{2}(\sigma_2 + \sigma_3) = \sigma_y, \text{ when } \sigma_2 \le \frac{1}{2}(\sigma_1 + \sigma_3) \tag{5.73a}$$

$$f' = \frac{1}{2}(\sigma_1 + \sigma_2) - \sigma_3 = \sigma_y, \text{ when } \sigma_2 \ge \frac{1}{2}(\sigma_1 + \sigma_3) \tag{5.73b}$$

When $\tau_y = 0.7\sigma_y$, we obtained parameter $b = 4/3$. The equations are

$$f = \sigma_1 - \frac{3}{7}(\frac{4}{3}\sigma_2 + \sigma_3) = \sigma_y, \text{ when } \sigma_2 \le \frac{1}{2}(\sigma_1 + \sigma_3) \tag{5.74a}$$

$$f' = \frac{3}{7}(\sigma_1 + \frac{4}{3}\sigma_2) - \sigma_3 = \sigma_y, \text{ when } \sigma_2 \ge \frac{1}{2}(\sigma_1 + \sigma_3) \tag{5.74b}$$

Therefore, the Yu-UYC encompasses the Tresca yield criterion and the Huber-Mises yield criterion as its special cases or linear approximations. In addition, a serial of new yield criteria can be introduced from the Yu-UYC. The application lets us use the accuracy and fine method for selecting the yield criterion other than use those two old criteria. Those two old criteria can only be used for materials with $t_y = 0.5\,\sigma_y$ or $\tau_y = 0.577\sigma_y$.

This method can be used for several examples in mechanics of materials and other related textbooks.

5.16 Summary

Yu-UST was proposed in 1991. Since then, it has been widely used in many fields in which a series of new results are obtained. The twin-shear unified strength theory won the National Natural Science Prize of China (NNSP) in 2011. The NNSP is the highest prize of science in China. Yu Mao-Hong won the Mathematic and Mechanics Prize of the Ho Leung Ho Lee Foundation for his contribution to strength theory in 2015.

Unified strength theory has been widely used in generalized plasticity, structural plasticity and computational plasticity and other fields, especially geomechanics and geotechnical engineering.

References

Fan, W., Yu, M.H. & Deng, L.S. (2018) *Strength Theory of Geotechnical Structure*. Science Press, Beijing. (in Chinese).

Ma, Z.Y., Liao, H.J. & Qi, J.L. (2017) *Numerical Analysis of Geo-Materials under Complex Stress*. Xi'an Jiaotong University Press, Xi'an. (in Chinese).

Wang, L.Z. & Zhang, Y.Q. (2011) Plastic collapse analysis of thin-walled pipes based on unified yield criterion. *International Journal of Mechanical Science*, 53, 348–354.

Yu, M.H. (2004) *Unified Strength Theory and Its Applications*. Springer, Berlin.

Yu, M.H. (2006) *Generalized Plasticity*. Springer, Berlin.

Yu, M.H. (2018) *Unified Strength Theory and Its Applications*, 2nd edition. Springer, Singapore.

Yu, M.H. & Li, J.C. (2012) *Computational Plasticity: With Emphasis on the Application of the Unified Strength Theory*. Springer and Zhejiang University Press, Hangzhou, China.

Yu, M.H., Ma, G.W. & Li, J.C. (2009) *Structural Plasticity: Limits, Shakedown and Dynamic Plastic Analyses of Structures*. Springer and Zhejiang University Press, Hangzhou, China.

Zheng, C.X. & Wen, Q. (2002) Explosive experiment of mild steel pressure vessel and research on formula of explosive stress. *Pressure Vessel*, 19(2), 9–12. (in Chinese).

Readings

[Readings 5–1] Some books about the applications of the Yu-UST are shown as follows:

1 *Unified Strength Theory and Its Applications* (Yu, 2004);
2 *Generalized Plasticity* (Yu, 2006);
3 *Structural Plasticity: Limits, Shakedown and Dynamic Plastic Analyses of Structures* (Yu et al., 2009);
4 *Computational Plasticity: with Emphasis on the Application of the Unified Strength Theory* (Yu and Li, 2012);
5 *Numerical Analysis of Geo-materials Under Complex Stress* (Ma et al., 2017), in Chinese;
6 *Strength Theory of Geotechnical Structure* (Fan et al., 2018), in Chinese;
7 *Unified Strength Theory and Its Applications*, second edition (Yu, 2018);
8 *Soil Mechanics: New Theory and New Results* (*Trilogy of Geomechanics* – Volume One), to be published;
9 *Rock Mechanics: New Theory and New Results* (*Trilogy of Geomechanics* – Volume Two), to be published;
10 *Concrete Mechanics: New Theory and New Results* (*Trilogy of Geo-mechanics* – Volume Three), to be published.

[Readings 5–2] Several papers about the applications of the Yu-UST are shown as follows:

11 Xu SQ, Yu MH (2005) Shakedown analysis of thick-walled cylinders subjected to internal pressure with the unified strength criterion. International Journal of Pressure *Vessels And Piping*, 82(9), 706–712.
12 Altenbach H, Kolupaev VA (2008) Remarks on Model of Mao-HongYu. In: The Eighth Int. Conference on Fundamentals of Fracture (ICFF VIII), Tong Yi Zhang, Biao Wang and Xi-Qiao Feng eds. 2008, 270–271.
13 Li J, Ma G, Yu MH (2008) Penetration analysis for geo-material based on unified strength criterion. International Journal of Impact Engineering, 35, 1154–1163.
14 Song L, Cho C, Lu S, et al. (2008). Study on softening constitutive model of soft rock using strain space based unified strength theory. International Journal of Modern Physics B, 22(31n32), 5375–5380.

15 Kolupaev VA, Altenbach H (2009) Application of the Unified Strength Theory of Mao-Hong Yu to Plastics. In: Tagung Deformations- und Bruchverhalten von Kunststoffen 24–26.06.2009, Book of Abstracts ed. by W. Grellmann, Merseburg, pp. 320–339 (in German)

16 Fan W, Yu MH, Deng L, et al. (2013). New strength formulae for rock surrounding a circular opening. Canadian Geotechnical Journal, 50(7), 735–743.

17 Ma ZY, Liao HJ, Dang FN (2013) Influence of Intermediate Principal Stress on the Bearing Capacity of Strip and Circular Footings. J. of Engineering Mechanics, ASCE, 140(7): 04014041:1–14.

18 Lin C, Li Y (2015). A return mapping algorithm for unified strength theory model. International Journal for Numerical Methods in Engineering, 104, 749–766.

19 Lin C, Sun Q, Chai Y, et al. (2017). Stress evolution in top coat of thermal barrier coatings by considering strength difference property in tension and compression. Surface and Coatings Technology, 329, 86–96.

20 Wang P, Qu S (2018) Analysis of ductile fracture by extended unified strength theory. International Journal of Plasticity, 104, 196–213.